Carbon fibres in engineering

Some McGraw-Hill books of related interest

Coull and Dykes: Fundamentals of Structural Theory
Zienkiewicz: The Finite Element Method in Engineering Science
Dugdale and Ruiz: Elasticity for Engineers
Penny and Marriott: Design for Creep
Bompas-Smith: Mechanical Survival: the use of reliability data

Carbon fibres in engineering

Editor: Marcus Langley

London . New York . St. Louis . San Francisco . Düsseldorf . Johannesburg
Kuala Lumpur . Mexico . Montreal . New Delhi . Panama . Paris . São Paulo
Singapore . Sydney . Toronto

Published by

McGRAW-HILL Book Company (UK) Limited

MAIDENHEAD . BERKSHIRE . ENGLAND

07 084421 6

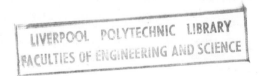
PRINTED AND BOUND IN GREAT BRITAIN

Contents

Contributors

Professor J. J. Bates, BSc, PhD, CEng, FIEE.
Head Electrical Engineering Branch, Royal Military College of Science, Shrivenham.

M. Bedwell, MA.
Lecturer in the Department of Mechanical Engineering, Lanchester Polytechnic, Rugby

W. G. Cook, BSc.
Rolls-Royce (1971) Limited.

A. G. Downhill, BSc(Eng), CEng, AFRAeS.
Rolls-Royce (1971) Limited.

Bryan Harris, PhD, BSc.
Reader in Materials Science, School of Applied Sciences, University of Sussex.

Marcus Langley, CEng, FIMechE, FRAeS.
Consulting Engineer.

M. Molyneux, API.
Technical Manager, Composite Materials Division, Fothergill & Harvey Limited.

R. Tetlow, MSc, CEng, AFRAeS.
Department of Aircraft Design, College of Aeronautics, Cranfield Institute of Technology.

Preface

The invention of high modulus, high strength carbon fibres in Great Britain was first disclosed in 1966 by the co-inventors, W. Watt, L. N. Phillips, and W. Johnson of the scientific staff of the Royal Aircraft Establishment, Farnborough.

The daily press and other popular media immediately headlined it as 'wonderful, ten times as strong as steel'. The new material certainly had remarkable properties which put it right ahead of conventional engineering materials in certain applications. But now, six years later, one may take a more realistic look at it and that is the object of this book. Although a great deal of research and development work remains to be done, certain practical applications may now be assessed in the light of some experience.

The initial price was high when production was still experimental and this restricted the use of carbon fibres to those cases where weight saving showed to the greatest advantage and where cost was of less importance than efficiency, as in some aerospace and weapon applications. Some of these are still 'classified' and must remain so, but the cost has fallen so considerably that engineers in more general fields are looking seriously at the material. Some of the examples in this book show how carbon fibres are starting to edge their way into structural, mechanical, and electrical usage.

We have attempted to describe the basic nature of carbon fibres and of the matrix materials associated with them, to guide design engineers in their use, to show how they may be handled on the workshop floor and to illustrate typical uses already in service. We feel that it is now possible to give some guidance in the development of new applications and to show where disappointments may be expected. We have also taken a sober look at future prospects.

Although the chapters have been written by experts in their respective fields, the language has been chosen so that the subject matter should be comprehensible not only to trained engineers but also to industrial managers and economists. We hope that it will be of value to students and to those responsible for their education.

We regret that military and commercial security have restricted the number of examples which could be given but we understand the reluctance to give publicity at this stage to projects which have not yet proved viable in service.

As Editor, I wish to express my thanks to the contributing authors who have been so helpful in their collaboration, and to Mr. E. Dellow who undertook considerable extra editorial work when I was incapacitated by an accident. I must also thank those other organizations and individuals who have provided additional information to add to the value of the book.

Marcus Langley

Reigate,
Surrey,
April 1972

Symbols and Abbreviations Used in the Text

Generally accepted chemical and physical symbols are not included

Greek

gamma	γ	Fracture energy
theta	θ	Angle of applied stress to fibre orientation
mu	μ	Poisson's ratio
nu	ν	Poisson's ratio
rho	ρ	Density, Specific gravity
sigma	σ	Tensile strength or stress
tau	τ	Shear strength or stress

Latin

E_*	(Young's) Modulus of elasticity
G_*	Shear Modulus; Modulus of rigidity
V_*	Volume fraction

* Subscripts c, f, and m refer to composites, fibres, and matrices

Acronyms

CFC	Carbon fibre composite(s)
CFRP	Carbon fibre reinforced plastic(s)
CFRTP	Carbon fibre reinforced thermoplastic(s)
DGEBA	Diglycidyl ether of bisphenol A (epoxy)
DMC	Dough moulding compounds
GRP	Glass reinforced plastic(s)
ILSS	Interlaminar shear stress
NDT	Non-destructive testing
NOL	US Naval Ordnance Laboratory ring test
PAN	Polyacrylonitrile
Prepeg	Pre-impregnated fibres, the matrix material being partially cured
SMC	Sheet moulding compounds
TAC	Thyristor assisted commutation
UTS	Ultimate tensile strength

SI and Metric Units

This list is not a complete set of conversion tables. It includes only those units found in the text and which are not in general use throughout the English-speaking world.

Centigrade (°C) temperatures are used throughout.

Symbol	SI or Metric Unit	Imperial Equivalent
m	Metre	39·3701 in
m^2	Square metre	10·7639 ft^2 or 1550 in^2
μ	Micrometre (10^{-6} m)	$10^{-6} \times 39\cdot3701$ in
Å	Ångström	$10^{-10} \times 39\cdot3701$ in
$MN\,m^{-2}$	Meganewton/m^2	145·05 lbf/in^2
$GN\,m^{-2}$	Giganewton/m^2	145 050 lbf/in^2 or 64·8 $tonsf/in^2$
$J\,m^{-2}$	Joule/m^2	0·068 54 ft lbf/ft^2
$kJ\,m^{-2}$	kilojoule/m^2	68·54 ft lbf/ft^2 or 0·476 ft lbf/in^2
kg	kilogramme	2·204 62 lb
Mg	Megagramme (tonne)	2204·62 lb or 0·984 207 ton

1. The Nature and Properties of Carbon Fibres and their Composites

Bryan Harris

1.1 Carbon Fibres

1.1.1 THE CARBON-CARBON BOND AND POLYMER CHAINS

From considerations of bonding in solids we expect that the strongest materials are likely to be those with a high density of covalent bonds and small, light atoms. We know that the carbon–carbon bond is strong, from spectroscopic studies of bond energies in compounds containing aliphatic C—C linkages, and the potential strength of carbon could be deduced from the fact that diamond is the hardest material known. Diamond is rather intractable and, costliness apart, is not normally considered an engineering material. But the C—C bond occurs in other familiar materials—the humbler and cheaper polymers. Polymers are not usually thought of as highly stable solids, but this is because in bulk form their behaviour is not determined by the C—C bond. The long-chain molecules, which are composed of many thousands of C—C bonds end to end, are bundled together in a random fashion, as shown in Fig. 1.1, being joined loosely here and there by weak inter-molecular forces. Heating or application of stress easily destroys these secondary bonds, even though the main-chain bonds remain intact. If the polymer is drawn into a fibre, the polymer chains are aligned and brought close together, as shown in Fig. 1.2. The polymer is denser and much stronger because the primary C—C bonds themselves are directly loaded. If the mechanical performance of a polyethylene fibre is assumed to be determined by the extension of the polymer backbone chain, such as that drawn schematically in Fig. 1.3, the theoretical elastic modulus and tensile strength of the fibre can easily be calculated. Based on an estimated 'cross-sectional area' for the polyethylene molecule of $18 \cdot 25 \times 10^{-20} \ m^2$ and spectroscopic data for the force constants for the stretching of the C—C bond and the opening of the bond angles under stress we find that Young's modulus should be

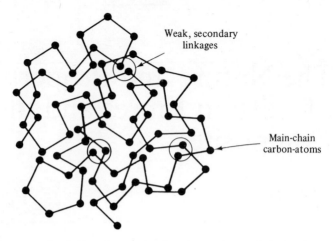

Fig. 1.1 *Structure of a bulk polymer, showing the randomly kinked main chain with occasional weak secondary linkages.*

about $250 \, \mathrm{GN \, m^{-2}}$.† If the chain is made rigid, however, by reducing the flexibility of the bond angle-opening mechanism, we obtain a value of E close to $1000 \, \mathrm{GN \, m^{-2}}$, which is about that of diamond. The theoretical strength of the fibre, estimated from measured bond energies, is of order 30 $\mathrm{GN \, m^{-2}}$. Such mechanical properties are never realized in polymers, but the figures give some notion of the incentive for the development of strong carbon fibres. Carbon fibre and cloth, made by pyrolysing cellulose, had been known since Edison first made lamp filaments from cotton. But these fibres were very weak, a mere ten or twenty times stronger than bulk graphites. At the other extreme Bacon[1] prepared graphite whiskers, small scroll-like bodies a

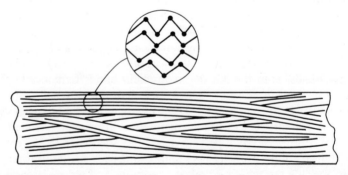

Fig. 1.2 *Structure of a drawn polymer fibre, showing the alignment of the main chains.*

† Note that Young's modulus for steel, about $30 \times 10^6 \, \mathrm{lb \, in^{-2}}$, is roughly $210 \, \mathrm{GN \, m^{-2}}$. It is usually convenient to use $\mathrm{GN \, m^{-2}}$ for modulus ($10^6 \, \mathrm{lb \, in^{-2}} \approx 7 \, \mathrm{GN \, m^{-2}}$) and $\mathrm{MN \, m^{-2}}$ for strengths ($10^3 \, \mathrm{lb \, in^{-2}} \approx 7 \, \mathrm{MN \, m^{-2}}$).

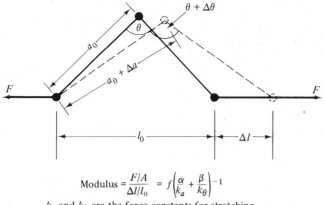

$$\text{Modulus} = \frac{F/A}{\Delta l/l_0} = f\left(\frac{\alpha}{k_a} + \frac{\beta}{k_\theta}\right)^{-1}$$

k_a and k_θ are the force constants for stretching
bonds and opening bond angles.

Fig. 1.3 *Model for the calculation of Young's modulus for a polymer fibre.*

few microns in diameter and up to 3 cm long, which had elastic moduli of
about $700 \, \text{GN m}^{-2}$ and strengths of $20 \, \text{GN m}^{-2}$. But whereas neither of
these materials was directly useful to the engineer, developments in manufacturing methods have resulted, as we shall see, in continuous carbon fibres
having properties close to those of whiskers, realizing to a much greater
extent than was first visualized the potential strength of the C—C bond.

1.1.2 PYROLYSIS OF TEXTILE FIBRES

One of the weaknesses of the polyethylene backbone chain is that the bond
angle distortion makes too great a contribution to the overall main chain
compliance. This weakness can be removed by modifying the main chain
configuration. For example, in polyphenylene oxide, aromatic rings are
linked by [—O—] bonds as shown in Fig. 1.4(a), and the rigidity of the main
chain is much increased. An alternative method is to use a structure known
for obvious reasons as a ladder polymer (Fig. 1.4(b)). Such a structure is
stronger because two bonds must be broken instead of one in order to sever
it. One of the most successful of the precursor polymers used for manufacturing carbon fibres has been polyacrylonitrile (PAN). As shown in Fig. 1.5(a),
PAN is not quite a ladder polymer, but it has the unique property of forming
a ladder structure when heated at about 200°C. The chemistry of PAN
pyrolysis has been clearly described by Watt.[2] The molecules in the precursor
fibre are highly oriented by stretching to some 30 times their original length
so that they resemble the polyethylene chains represented in Fig. 1.2. As
Fig. 1.5(a) shows, the polymer is atactic—that is to say the nitrile (CN) groups
which substitute for an H group at alternate points in the polyethylene chain
are randomly arranged on either side of the main chain. But where several

3

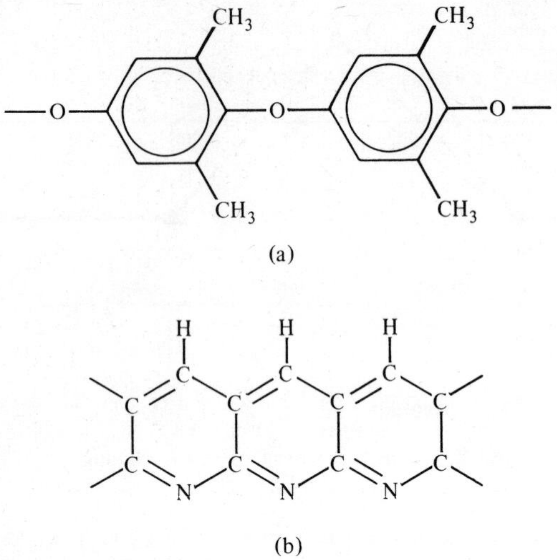

(a)

(b)

Fig. 1.4 *(a) Structure of polyphenylene oxide, (b) Structure of a ladder-type polymer.*

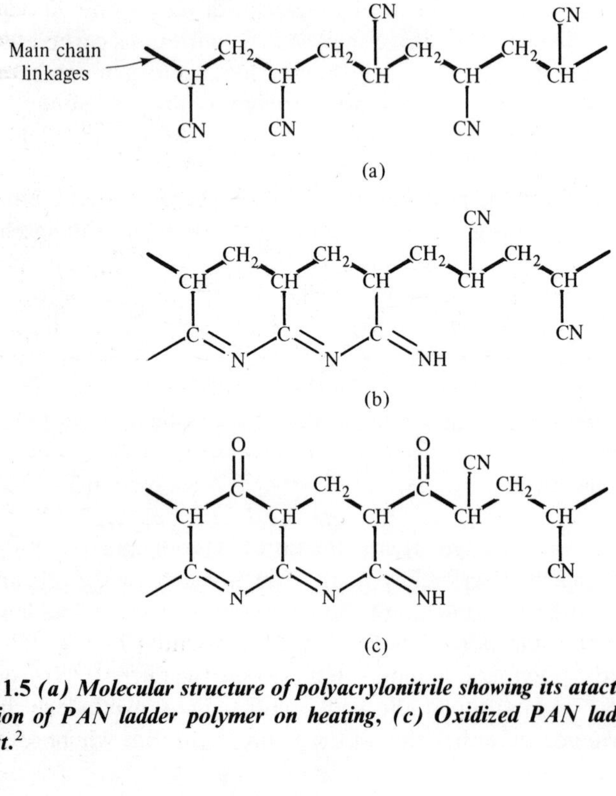

Main chain linkages →

(a)

(b)

(c)

Fig. 1.5 *(a) Molecular structure of polyacrylonitrile showing its atactic nature, (b) Formation of PAN ladder polymer on heating, (c) Oxidized PAN ladder polymer, after Watt.*[2]

4

CN groups occur in sequence on the same side of the chain, heating at about 200°C encourages the formation of a ladder structure (Fig. 1.5(b)) which renders the polymer thermally stable and prevents melting during the subsequent carbonization process. The PAN fibres are in fact heated in air at 220°C when, in addition to ladder polymer formation, some of the CH_2 groups are oxidized, as shown in Fig. 1.5(c). During subsequent heating at higher temperatures in an inert atmosphere additional cross-linking reactions occur between adjacent chains with elimination of H_2O, HCN, and N_2, with the result that ribbons are gradually built up which consist largely of carbon atoms arranged in aromatic ring structures. These ribbons are now beginning to resemble the basal planes of true graphitic structures although the distances between them are much greater than in graphite and the atoms in the planes are not ordered with respect to those in neighbouring planes (a turbostratic structure).

Once the evolution of gases has diminished, heating can be continued to temperatures above 400°C without disintegration, and the strength and modulus of the fibres increases rapidly. At 1000°C the fibres have lost some 50 per cent of their original weight. The atactic structure of the precursor polymer, and the nature of the drawing process result in misalignment of the sheets of carbon atoms with respect to the fibre axis and considerable imperfection of the structure within the planes. To improve the degree of perfection and, in consequence, the mechanical properties, the fibres are prevented from shrinking or are actually stretched during pyrolysis. If the carbonized fibres are heated at temperatures above 2000°C the degree of perfection of the carbon increases and the structure becomes more representative of true graphite, but while this increased perfection results in a higher elastic modulus, the tensile strength is reduced. Commercial fibres are usually offered in the 'carbonized' form, with low modulus and high strength, or in the 'graphitized' form, with high modulus and lower strength. Figure 1.6 shows the relationship between strength and modulus of PAN fibres as a function of heat treatment temperature. Additional strengthening can be obtained by hot stretching fibres above 2000°C when they can undergo plastic deformation. This process of strain-graphitization has been described by Johnson et al.,[3] who show that the degree of preferred orientation, the fibre strength, and the elastic modulus all increase markedly during stretching (Fig. 1.7). Increases in modulus from 420 to 670 GN m^{-2} have been obtained by extensions of 30 per cent at 2750°C and it is interesting to note that whereas the modulus of hot-stretched fibre approaches the crude theoretical estimate, the fibre strength is still an order of magnitude lower.

The technical procedure for the manufacture of carbon fibres is roughly as follows. First, in the textile part of the process, the PAN is wet-spun—that is a solution of the polymer is forced through a spinneret containing a series of minute holes into a stabilizing bath where the polymer threads are

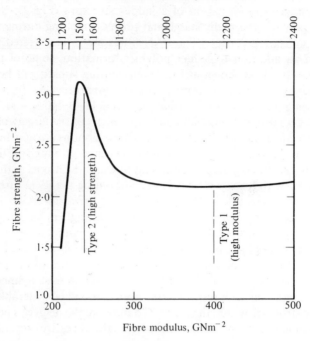

Fig. 1.6 *Dependence of strength and modulus of carbon fibres on heat-treatment temperature, after Watt.*[2]

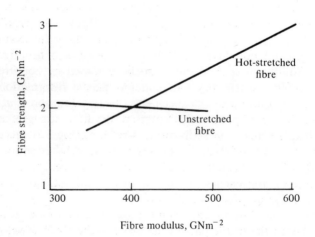

Fig. 1.7 *Effect of hot-stretching on the strength and modulus of PAN carbon fibres (after Johnson et al.*[3]*).*

6

first formed. These threads are dried and stretched to some ten times their original length at a slightly elevated temperature so as to draw out and align the main polymer chains and increase interchain adhesion. The precursor polymer is usually produced from the spinneret in tows of some 10 000 filaments. These filaments are oxidized by heating in air, and the first stage of ladder polymer formation occurs. During this step the fibres must be restrained from reverting to their weak, unstretched state by preventing them from shrinking (or even by stretching them slightly). After a few hours heating in air at 220°C the fibres are black and the oxidized structure, which is quite rigid, has no further tendency to shrink. The fibres are then carbonized by heating at a controlled rate to 1000°C in an inert atmosphere, and after gas evolution has ceased heat treatment is continued at higher temperatures until the required properties are obtained. The early experimental manu-facture of carbon fibres was carried out by batch methods, producing metre length tows, but more sophisticated plant is now used to produce semi-continuous lengths (1000 m or more). This has the added advantage that subsequent fibre surface treatments, or the filament-winding of composite structures, or the vapour deposition of metal coatings on the fibres, can also be carried out semi-continuously.

Fibres manufactured by European companies are usually made from acrylic polymers. PAN-based precursors, such as the Courtelle-type material favoured by Morganite Modmor, Courtaulds, and other British establish-ments, have so far proved to be the most satisfactory thermoplastic precur-sors, but cellulosic materials such as Rayon have been used successfully by manufacturers in the US. Rayon, like polyphenylene oxide, does not have an all-carbon backbone chain; each monomer unit contains an oxygen atom, and this carbon–oxygen link, which is relatively unstable, must be eliminated during carbonization. During slow pyrolysis in an inert atmosphere the rayon undergoes a complex series of reactions, including removal of water, main chain scission, and the formation of an intermediate compound, laevoglucosan, which subsequently decomposes to form a carbonaceous char. The tensile properties of rayons pyrolysed at temperatures up to 1000°C are not high, and further treatment, including a hot-stretching process, is needed to develop properties as good as those of PAN fibres. However, in order to achieve moduli as high as $700 \, \text{GN m}^{-2}$ extensions of the order of 300 per cent are needed at 2750°C. Such extensions have not been possible with PAN fibres. Fibres made from cellulose precursors are very similar in terms of structure and properties to those produced from PAN although, unlike PAN fibres, they have irregular cross-sectional shapes.

1.1.3 CARBON FIBRES FROM NON-TEXTILE PRECURSORS

Textile-based carbon fibres have so far proved to be expensive and estimates of future cost, based on quantity production, do not fail much below about

7

£10/kg. Consequently, the possibility of producing much cheaper fibres from low-cost raw materials such as resins, hydrocarbon pitches, or lignin pitches is very attractive. Thermo-setting resins such as phenol-formaldehyde and phenol-hexamine carbonize to form glassy carbon—a form of carbon that is highly resistant to graphitization because of the three-dimensional cross-linked structure of the original resin. Jenkins[4] has described experiments in which melt-extruded phenol-hexamine is carbonized to glassy fibre with strengths as high as those of some PAN fibres but with lower elastic moduli more characteristic of glass or silica. This lower rigidity is a considerable drawback, since it is principally on account of its high elastic modulus that carbon is an important rival to glass fibre.

Crude hydrocarbon pitches can be refined by distillation and melt-spun into fibres. Early experiments on the controlled oxidation and carbonization of such fibres yielded carbon filaments, which were considered to have a glassy structure, with properties rather worse than those of textile-based fibres. More recently, however, Hawthorne[5] has hot-stretched such fibres between 2000°C and 2500°C and has found that strain-graphitization occurs which results in the development of a highly oriented structure. Elastic moduli as high as 600 GN m^{-2} can be produced by stretching some 180 per cent, and the structures of these fibres are similar to those made from textiles. Figure 1.8 shows the development of fibre anisotropy during hot-stretching.

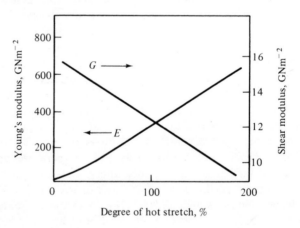

Fig. 1.8 *Effect of hot stretching on the elastic moduli of pitch-based carbon fibre (after Hawthorne[5]).*

The tensile strength/modulus relationship for these fibres is similar to that shown in Fig. 1.7, and Hawthorne reports strengths as high as 2·6 GN m^{-2}, but the process is not yet sufficiently advanced to prevent large batch-to-batch variations in strength.

There have been many studies of the fine structure of fibres obtained from various precursors, and although the proposed models differ in detail, there is general agreement on the nature of the principal structural features that determine fibre properties. Most types of filament have a fibrillar structure, developed from that of the polymer fibre, in which the fibrils themselves are composed of stacks of graphitic planes. The two principal models are those of Ruland,[6] developed for rayon-based fibres, and of Johnson,[7] for PAN fibres. These models are illustrated in Fig. 1.9. Johnson's model suggests

L_c

(a) Johnson and Tyson Model, showing sub-grain boundaries, interfibril boundaries, and voids where crystallites are imperfectly in contact.

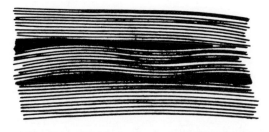

(b) Ruland Model, showing bent ribbons and voids

Fig. 1.9 *Schematic representation of the models of (a) Johnson and Tyson[7] and (b) Ruland and coworkers[6] for carbon fibres, based on the interpretation of x-ray diffraction results.*

small blocks of crystal stacked end to end with crystallite boundaries within the fibrils at which small changes in orientation occur. Ruland's model represents the fibrils as continuous, bent ribbons in which the stacks of planes extend over long distances along the fibre length. Both models postulate oriented voids between the fibrils, and the volume fraction of voids is much larger in the rayon-based material. The average stack dimension, L_c, is roughly the same in the two models, some 20 to 100 Å, depending on heat-treatment temperature. Both models explain the observed dependence of

9

strength and modulus on heat-treatment conditions. The pores, which result from misfit between crystallites, are of various sizes in the initial structure. But at higher heat-treatment temperatures as the crystallite size increases the pore size also increases, while the void volume fraction actually falls. The degree of preferred orientation and the perfection of the crystal planes controls the elastic modulus, which therefore increases as the heat-treatment temperature rises. But if the fibre strength is controlled by the size of the sharp edged pores, it must fall, as observed, when the pore size increases. On the other hand, strain graphitization tends to pull the fibrils straight, so that as well as increasing the degree of orientation, stretching closes up the pores and the strength increases. Ruland and his co-workers have shown that the strong dependence of modulus on orientation which is observed experimentally for fibres of many different types is accurately predicted by the theoretical model, as shown in Fig. 1.10. The elastic unwrinkling of the fibrils in this

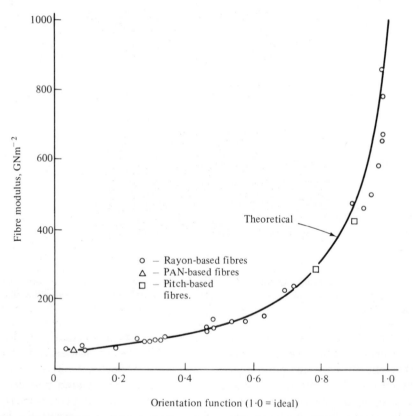

Fig. 1.10 *Comparison of the predictions of Ruland's model for the orientation-dependence of the elastic modulus with experimental results obtained for fibres of various types (after Fourdeux et al.[6]).*

10

model also accounts for the observation that the fibre modulus increases with strain.

The strength of carbon fibres is not catastrophically reduced by normal handling and in this respect they are easier to use than glass fibres which must be protected by sizing immediately after manufacture. Although this implies that the strength of the fibres is not sensitive to surface flaws it has been shown, nonetheless, that the strength can be increased to a certain extent by polishing treatments. Johnson[8] has shown, for example, that when fibres pyrolysed between 500 and 900°C are polished by controlled surface oxidation their strengths are higher than those of unpolished fibres (Fig. 1.11),

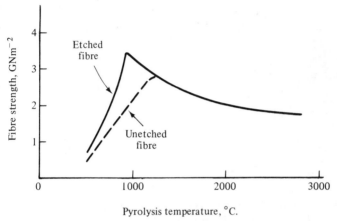

Fig. 1.11 *Effect of polishing on the strength of PAN carbon fibres, after Johnson.*[8]

which indicates that the strength of the fibres is controlled by surface flaws. Above 900°C the strength of polished fibres begins to decrease, and above 1250°C the strength is the same as that of unpolished fibres, which implies that defects other than surface flaws are then governing the strength.

1.1.5 RIVALS TO CARBON FIBRES

The chief rivals to carbon as a high strength, high modulus reinforcing fibre are the humble glass, the more exotic boron, and whiskers of such inherently strong solids as silicon carbide or aluminium oxide. Some of the mechanical properties of these materials are compared in table 1.1. It is easier to assess the relative merits of the continuous fibres, glass, carbon, and boron separately, and then to consider the special problems of short fibres. First, in terms of cost it is clear that glass, at a few pence per kilogram, has a distinct advantage over both carbon, at some £50/kg, and boron at three or four times that price. Future estimates of around £5/kg for carbon, based on production units of 2000 tonnes/year, still do not appeal to potential users. Large-scale

11

Table 1.1 *Properties of some reinforcing fibres*

Material	Manufacturer	Diameter, μm	Density, $10^3 \, kg\,m^{-3}$	Strength, $GN\,m^{-2}$	Modulus, $GN\,m^{-2}$†
Carbon fibre					
Modmor 1	Morganite	7·5	2·0	1·7–2·1	385–415
Grafil-HM	Courtaulds				
Carbon					
Modmor 2	Morganite	8·0	1·7	2·4–2·8	207
Grafil HT	Courtaulds				
Carbon					
Rigilor AC	Le Carbone	12·4	1·7–1·8	1·9–2·2	170–180
Rigilor AG	Le Carbone	11·0	2·1	0·8–1·0	250–300
Carbon					
Thornel 25	Union Carbide	6·6	1·5	1·4	175
Thornel 50	Union Carbide	6·6	1·63	2·0	350
Boron fibre	Avco	100	2·5	3·5	420
SiC whiskers	ERDE	0·5–2·0	3·15	6–10	450
S-glass	Fibreglass	about 10	2·49	up to 4·6	84
E-glass		about 10	2·55	1·8–3·5	72

†Note 1 million $lb\,in^{-2} \approx 7\,GN\,m^{-2}$.

production of boron fibre may not even be feasible; and there are certainly no predictions better than about £40/kg by the late 'seventies. Although the strengths of carbon and boron are no higher than that of glass, they are both some six times more rigid, and it is this advantage which is their principal claim to superiority over glass. But whereas most glasses lose their strength at relatively low temperatures as they approach the glass transition, carbon and boron retain theirs to much higher temperatures. Indeed carbon is the only material known that gets stronger as the temperature is raised. On the other hand carbon and boron are both susceptible to severe oxidation if exposed to air at temperatures above 400°C, whereas glass is unaffected, being itself a highly stable oxide. Boron and type 2 carbon fibres both have microcrystalline structures: in contact with certain active elements (nickel, for example, in the case of carbon) there is a possibility that changes in fibre structure will occur, with disastrous consequences to the fibre strength. Glass does not suffer in this respect. All three materials are produced as continuous fibres, and can therefore be used in filament winding methods of manufacturing composites that permit maximum utilization of their good mechanical properties. Both boron and glass fibres must be protected from abrasive damage, however. At the present time the manufacturing technology for boron/epoxy and boron/aluminium composites is at least as well developed in the US as carbon fibre composite technology is in the UK, and testing of structural components for aerospace applications in both types of materials is well advanced. For reinforcement of metals, however, it is perhaps not plain boron, but boron fibre coated with silicon carbide (Borsic) that will be the most serious rival to carbon. It has been shown that aluminium/Borsic composites have good fatigue properties, good resistance to corrosion, and can be fluxless-brazed. The specific modulus (E/ρ) in the fibre direction is comparable with those of boron/epoxy, carbon/epoxy, and titanium, but, more important, the lateral and torsional stiffnesses are some ten times greater. Furthermore, the fibres themselves are not oxidized when heated in air.

As table 1.1 shows, good quality whiskers have better mechanical properties than any of the continuous fibres. Because of the manner of their growth, however, they occur in the form of very fine, short crystals and they cannot therefore be handled by the techniques used with continuous fibres. Nevertheless, means have been developed for grading, aligning, and forming them into composites, with both resin and metal matrices, that have excellent mechanical properties.[9] Because of the short lengths of whiskers, and the fact that alignment is rarely as good as it is in composites containing continuous fibres, the full potential strength of whisker composites cannot easily be realized. But the mechanical properties of whiskers are so superior to those of other fibres that there is ample tolerance for losses in the composite fabrication process. The cost of whiskers is very much higher at present than that of carbon fibre, partly because they are manufactured by batch methods and partly because there is not yet even the small market enjoyed by carbon fibre.

1.1.6 PROSPECTS FOR THE FUTURE

At the present time it is not altogether clear whether carbon fibres have as commanding an advantage over their rivals as manufacturers would like. Neither is it clear to designers which of these new materials is likely to be first to overcome development difficulties and become acceptable as a sound engineering material. The question of cost has always obscured the problem of selection for manufacturers who are unwilling to pay the high price for what seem at the moment to be marginal advantages. In the one rather exclusive case where the superior material properties and the high overall cost of the product—the Rolls Royce RB 211 engine—more than outweigh the relatively minor consideration of fibre cost, the enterprise has failed, but for other reasons, and carbon fibre reinforced plastics have unjustly gained a bad reputation. These setbacks can be overcome, although perhaps not with the present state of the carbon fibre art.

What are the future prospects for utilization of carbon fibres both in specialist applications, such as in aerospace where costs tend to be secondary to performance, and in more down-to-earth engineering fields, where cost must always be a primary consideration? We shall consider the possibilities either for improving the mechanical performance or for reducing the cost of the fibre.

(a) *The fibre diameter.* As will be shown later there are reasons why it could be useful to have carbon fibres of larger diameter than those now available. This is partly because the fracture toughness of composites increases with the diameter of the reinforcing fibre, and partly because it eases somewhat the task of producing aligned composites. The present standardization in size has come about presumably because the precursor is a conventional textile fibre produced by textile methods, and the usual spinneret size yields a carbon fibre of about 8 μm in diameter. To produce thicker fibres it would be necessary to scale up the whole process, and there may be some difficulty in obtaining the degree of stretching during drawing and pyrolysis that are essential if a high degree of molecular and crystallite orientation are to be obtained. Figure 1.12 shows how strongly the mechanical properties of PAN-based fibres depend on diameter[10] and this suggests that the prospects for

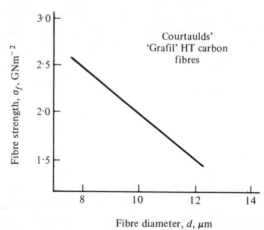

Fig. 1.12 *Strength of PAN carbon fibre as a function of fibre diameter, after de Lamotte et al.*[10]

achieving adequate properties in, say, a 30 μm fibre, are not encouraging. On the other hand, fibres produced from pitch do not undergo the textile manufacturing part of the process and even though they must be hot stretched by some 200 per cent Hawthorne has shown that it is possible to produce very strong, high modulus fibres with diameters as large as 30 μm. Time will tell whether these fibres have not also the added advantage of cheapness.

A recent report from NASA[54] describes the properties of large diameter carbon filaments (0.05–0.25 mm) manufactured by pyrolytic deposition.

14

These fibres are large enough to permit the use of the same fabrication techniques as are used in boron composite work, which makes for easy handling. Strengths of the order of 1 GN m^{-2} have been obtained with elastic moduli as high as 240 GN m^{-2}, and these values are substantially increased if a boron-containing gas is used for the process. These fibres may prove to be serious competitors to textile-precursor fibres if they can be manufactured in sufficiently large quantities.

(b) *Fibre strengthening*. Although commercial fibres are usually not of the highest available rigidity, it is nevertheless possible to obtain moduli close to the theoretical value for the C—C bond, given that the bond angle opening mechanism is somewhat restricted by the aromatic ring structure. Therefore although there seems to be limited scope for improving the elastic modulus, the fibre strength still falls well below expectation. The search for fibres with improved properties seems to be concentrated at present in experiments to find better precursor polymers or more efficient means of orienting the molecular structure prior to pyrolysis. Indeed both of these objectives could be achieved at once if a method could be found of making the precursor polymer isotactic instead of atactic. This would allow it to crystallize and the packing of chains in the drawn fibre would be much closer. Specific catalysts are known that can do precisely this in polystyrene, which is also normally atactic. Moreton[11] has recently shown that stretching wet-spun PAN fibres in steam and at temperatures up to 160°C can increase the strength and modulus of the final carbon fibre, and that polymers of high molecular weight could be stretched further than ordinary polymers, again resulting in better properties.

Other experiments have been aimed at improving the properties of the carbon fibres themselves. For example irradiation of type 1 and type 2 fibres to neutron doses of about 2.2×10^{17} n cm^{-2} increases both the strength and modulus by some 10 per cent.[12] Doping with boron to about 1 per cent also increases the modulus markedly (Fig. 1.13) and although the strength is not increased it seems likely, nonetheless, that such methods, perhaps in combination with fibre polishing treatments, could give substantial benefits.

(c) *Long and short fibres*. To make the best use of high strength and rigidity, the fibres must be accurately aligned by methods such as filament winding and electroforming, or by careful laying-up of preimpregnated sheets of fibre. The result is a highly anisotropic material with poor transverse properties. Admittedly a certain degree of cross-bracing, such as that used to confer adequate torsional stiffness in the RB 211 fan blade, does not seriously reduce the longitudinal properties, but any attempt to introduce even two-dimensional isotropy results in the loss of some 60 per cent of the maximum available strength. Clearly it is not sensible to use expensive, high quality fibres for such purposes: but there are several possibilities for the production of

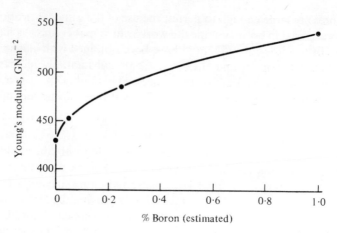

Fig. 1.13 *The effect of doping with boron on the modulus of carbon fibre, after Allen et al.*[12]

much cheaper fibres which could be used in this way, especially if they could be used in chopped form, for example by manufacturing fibre in the form of heavy tows, or by using a staple fibre (i.e., spun, like cotton) as a precursor, or by using scrap fibre that can be chopped and reused. These short fibres can probably not be processed into composites by simply blending with metal or plastic powder and extruding or pressing, because the fibres would be severely damaged and perhaps made shorter than the acceptable minimum length for reinforcement. But liquid infiltration methods, in combination with the whisker handling techniques discussed earlier, are certainly capable of giving very valuable materials.[9] In such systems the fibre/matrix bond strength is likely to be a much more important parameter than it is in continuous fibre composites.

1.2 Matrix Materials for Composites

Carbon fibres can be satisfactorily used only if they are embodied in a composite material. The matrix serves the dual purposes, among others, of protecting the fibres from damage and of transferring stress into them. The more efficiently it can do this the better use can be made of the fibre properties. Five types of material have so far been used as matrices in composites; metals, plastics, glasses, ceramics, and other forms of carbon. The questions that we ask in this section are, 'What are the present limitations on existing engineering materials falling within these five groups?' and, 'What advantages could the carbon fibre reinforced matrices have over the best monolithic engineering materials?' After discussing the mechanics of composites in section 1.3, we shall consider some of the developments and limitations in composites based on these diverse matrix materials.

16

Metals are the most versatile of engineering materials, and they owe this versatility to the fact that they can be plastically deformed and strengthened by a variety of methods which by and large act by inhibiting the motion of dislocations. Ironically, as a consequence of the non-directional nature of the metallic bond dislocations are highly mobile in pure metals which are therefore very soft. However, by exercising careful control over the introduction and distribution of defects the materials scientist can to a large extent tailor the properties of a metal or alloy system to suit his requirements. There are limitations, however. Increases in strength can usually be achieved only at the expense of the capacity to deform plastically, with the consequence that the strongest alloys lack the vital quality of toughness. Since brittleness is a drawback that no designer can afford to underestimate, this leads to the use of large safety factors which in turn means that the full potential of high strength alloys can frequently not be realized in practice. Brittle materials lack the advantage of being fail-safe that accompanies the virtue of toughness.

Many of the solid-state hardening methods used in alloys involve producing a material in a metastable state which subsequently tends to revert to the stable but non-strengthened condition if sufficient driving force is provided. Thus alloys strengthened by precipitation hardening, such as strong aluminium alloys, those depending on phase transformations of the martensitic type, such as steels, or even those depending simply on the presence of a high dislocation density, as in heavily cold-worked materials— these will all tend to soften at elevated temperatures. The strongest aluminium alloys begin to lose their strength at temperatures little over 150°C, for example. Furthermore in metastable alloys the problem of fatigue is intensified, because the cyclic deformation process generates large quantities of point defects which can cause local reversion of the metastable alloy structures, even at room temperature.

Many conventional metallic systems have the disadvantage of being relatively heavy. This may not be significant in land-based engineering projects, but it is of serious consequence in aerospace engineering. As Biggs[13] has shown, the greatest gains in the materials field are to be had from reductions in density rather than from increases in strength and rigidity, particularly where the predominant design loads are compressive, as in columns and struts.

In these three areas, the reinforcement of metals with carbon fibres could provide materials with advantages over conventional alloy systems. They cannot, however, be expected to provide materials that can compete on a straightforward strength-at-room-temperature basis with, say, the best steels, especially if the cost factor is included in the reckoning.

17

1.2.2 POLYMERIC MATERIALS

Few polymers are thermally stable by comparison with metals or ceramics, and even the most stable, like polyimides and polybenzoxazoles, are degraded by exposure to temperatures above about 300°C. There is little that reinforcement can do about the chemical aspects of degradation, but the associated fall in strength and increase in creep deformation—a feature common to all polymers, though less serious in cross-linked resin systems than in thermoplastics—can indeed be prevented by fibre reinforcement. A more serious problem in polymers is their very low mechanical strength in bulk form; and like metals, the weakest polymers are tough but the strongest tend to be brittle. In such materials there is the greatest scope for improvement and it is in the field of fibre reinforced plastics that the greatest successes have so far been achieved. Polymers are traditionally insulators, and in their application as such their strength is perhaps a secondary consideration. It is conceivable, however, that the enhanced electrical and thermal conductivity of plastics reinforced with carbon fibres could be beneficial in certain applications. Most polymers are already low density materials, and the addition of fibres cannot confer any advantages in this respect.

1.2.3 GLASSES

Glasses have high chemical stability, but many lose their mechanical strength at relatively low temperatures as they pass through the glass transition. Special glasses have been developed with high transition temperatures, however, and many are at least as resistant as some of the less stable steels. The principal problem with glassy materials is that they are always brittle, and are therefore regarded with suspicion by designers. Design stresses must be predominantly compressive. The difficulty is not that glasses are not strong, for with careful preparation very high strengths can be achieved. The drawback is that they are highly notch-sensitive and their measured strengths are in consequence subject to wide variation. Any reported value is the mean of a large number of test results, but this will be an unreliable indication for design purposes since the standard deviation of the results will invariably be great. Glasses are not able to relieve stress concentrations at crack tips by plastic deformation in the way that even quite strong steels can, and they are not fail-safe. A concomitant disadvantage, aggravated by low thermal conductivity, is that glass usually has poor thermal shock resistance unless, like quartz, it also has a low thermal expansion coefficient. Many of these difficulties can be overcome by reinforcement with carbon fibres, and with the bonus of a saving in weight.

1.2.4 CERAMICS

Most ceramics, by which we mean conventional whiteware and porcelains as well as the more exotic pure ceramics of the MgO or Al_2O_3 variety, suffer from the same defects as glasses in the sense that though they are potentially high strength solids, they are also brittle and highly notch-sensitive. Most ceramics retain their strength to very high temperatures, however, unlike most glasses, and many have excellent thermal shock resistance. Silicon nitride is an example of a ceramic that could be a valuable engineering material were it not for its brittleness. Improving toughness and reducing notch sensitivity are perhaps the only reasons for attempting to reinforce such materials, for as Bowen[14] points out the modulus of many ceramics is not very different from that of carbon fibre, and there is little point in attempting to reinforce any except those having low rigidity and strength.

Concrete is a ceramic material in which crystalline aggregate particles are embedded in a pseudo-glassy silicate matrix. There is much interest in the possibilities of reinforcing it with carbon fibres but at the time of writing very little information is available on the results of experiments. It seems unlikely that in the vast tonnages consumed in structural building practice sufficient carbon fibre could be incorporated by normal mixing methods to make economically feasible improvements in strength or toughness, compared with what can be achieved with glass fibre or steel wire for example. But for special applications it may be possible to produce high volume fraction composites containing aligned fibres by such methods as slip casting or vacuum dehydration of a slurry in combination with filament winding. Thin-walled pipes of high bursting strength could be made in this way.

1.2.5 CARBON

Carbon has many attractive engineering qualities. It can be prepared in a variety of forms—conventional hot pressed carbons and graphites, densified (impermeable) graphite, pyrolitic graphite, and vitreous carbon (for a discussion of the properties of these forms, see Cahn and Harris[15])—with a wide range of engineering properties. It is valuable for its lubricating properties, its electrical conductivity (and electrical resistance!), its nuclear properties and, in the pyrolytic and vitreous forms, for high strength and resistance to oxidative and chemical attack. The opportunity to improve the mechanical properties of such an important material and reduce its brittleness somewhat has been the driving force behind several attempts to produce what, for aerospace and rocketry purposes, would be an invaluable development— carbon reinforced carbon.

1.3 Mechanics of Composite Strengthening

The general principles of composite strengthening will be applied here to the specific case of carbon fibre reinforcement, that is for strengthening with brittle fibres.

The elastic modulus of a composite consisting of continuous, aligned fibres in a softer matrix is given by simple elasticity theory

$$E_c = E_f V_f + E_m(1 - V_f) \tag{1.1}$$

where the subscripts c, f, and m refer to composite, fibre, and matrix, and V_f is the fibre volume fraction. This expression is strictly true only for the case where both fibre and matrix are isotropic and it assumes that the two components are constrained to deform together. This elementary 'Rule of Mixtures' is in fact a lower bound, and will generally be exceeded by an amount depending upon the difference in Poisson's ratio of the two constituents as a result of the lateral elastic constraint which the two phases exert on one another. The equivalent theoretical expression for the shear modulus[16] is

$$G_c = G_m \left[\frac{(G_f + G_m) + (G_f - G_m)V_f}{(G_f + G_m) - (G_f - G_m)V_f} \right] \tag{1.2}$$

and the variation of G_c with V_f for various values of the stiffness ratio, G_f/G_m, is shown in Fig. 1.14 (after Tsai, Adams, and Doner[17]).

If the fibres are not continuous, an allowance must be made for the fact that each short fibre is unable to carry a stress near its ends, loads being transmitted into the fibres by shear along the fibre/matrix interface. However,

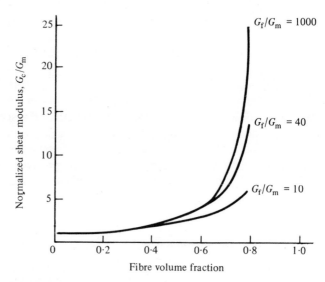

Fig. 1.14 *Theoretical variation of the composite shear modulus, G_c, with fibre volume fraction, after Tsai et al.*[17]

the influence of this ineffective length is small if the fibres are longer than about 50 times the diameter.

An aligned composite is highly anisotropic, and the three familiar elastic constants, E, G, and v for isotropic materials are insufficient to characterize the elastic behaviour of such materials. In an aligned composite it is usually possible to assume that the material is transversely isotropic, however, and if the material is being used in plate form, as is frequently the case with laminates, the assumption of a plane stress state reduces the number of elastic constants in the Hooke's law matrix to four:

$$
\begin{bmatrix} \varepsilon_1 \\ \varepsilon_2 \\ \gamma_{12} \end{bmatrix} = \begin{bmatrix} S_{11} & S_{12} & 0 \\ S_{12} & S_{22} & 0 \\ 0 & 0 & S_{66} \end{bmatrix} \begin{bmatrix} \sigma_1 \\ \sigma_2 \\ \tau_{12} \end{bmatrix} \tag{1.3}
$$

where, with reference to Fig. 1.15, the compliances are:

$$S_{11} = 1/E_{11}$$
$$S_{22} = 1/E_{22}$$
$$S_{12} = -v_{12}/E_{11} = -v_{21}/E_{22}$$
$$S_{66} = 1/G_{12}$$

The variation of elastic properties with orientation can be found by transforming the elasticity matrix. For example the variation of E with the angle, θ, between fibres and applied stress is given by

$$1/E_\theta = S_{11} \cos^4 \theta + S_{22} \sin^4 \theta + (S_{66} - 2S_{12}) \cos^2 \theta \sin^2 \theta \tag{1.4}$$

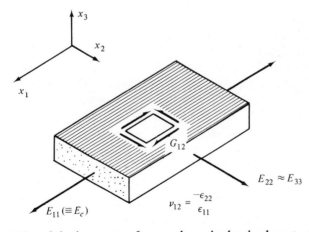

Fig. 1.15 *Definition of elastic constants for an orthotropic plate in plane stress.*

21

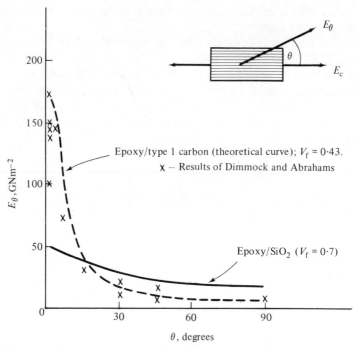

Fig. 1.16 *Comparison of experimentally measured values of modulus, as a function of fibre orientation, with theoretical predictions, after Dimmock and Abrahams.*[18]

Figure 1.16 shows some results of Dimmock and Abrahams[18] in which the theoretical elastic modulus of carbon fibre reinforced epoxy resin ($V_f = 0.43$), as a function of fibre orientation, is compared with experimental values. The agreement is good despite the fact that the elastic theory strictly applies only to composites containing isotropic fibres. E_θ falls off rapidly as θ increases, but considerable benefits may be obtained by cross laminating as shown in Fig. 1.17 where the orientation-dependence of E and G in Rolls Royce carbon fibre composites (Hyfil) is compared for aligned material and for laminates built up from equal numbers of plies at angle $\pm\theta$ to the applied stress.[19]

Two-dimensional isotropy can be obtained in a composite by randomizing the orientations of the fibres in the plane of the plate. The modulus is found by averaging E_θ over all values of θ:

$$\langle E_\theta \rangle = \frac{\int_0^{\pi/2} E_\theta \, d\theta}{\int_0^{\pi/2} d\theta} \tag{1.5}$$

The solution of eq. (1.5) for reinforced resins and several values of the modulus ratio E_f/E_m is shown in Fig. 1.18.[20] Clearly for CFRP, where this ratio can be as high as 200, the loss of rigidity would be catastrophic.

22

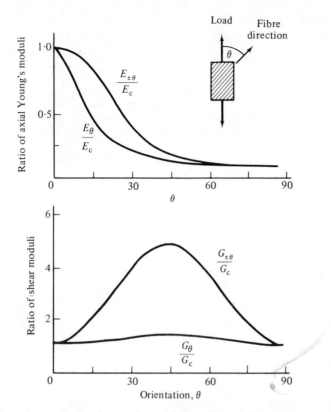

Fig. 1.17 *Effect of cross-laminating on the axial Young's moduli and shear moduli of carbon fibre composites, compared with results for unidirectional composites, after Goatham.*[19]

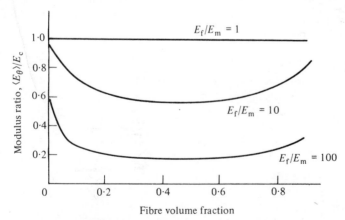

Fig. 1.18 *The effect of randomizing the fibre orientation, as a function of V_f, for various values of the modulus ratio, E_f/E_m, after Nielsen and Chen.*[20]

The tensile strength of a composite containing more than a certain minimum volume fraction of aligned, continuous, brittle fibres is given, like the modulus, by a rule of mixtures:

$$\sigma_c = \sigma_f V_f + \sigma'_m(1 - V_f), \qquad (V_f < V_{min}) \qquad (1.6)$$

where σ'_m is the tensile stress in the matrix when the composite is strained to its ultimate stress and σ_f is the fracture stress of the fibre. When the fibres are not continuous, the stress in the fibres is not uniform but builds up from the ends, as shown schematically in Fig. 1.19. If the fibre is shorter than a

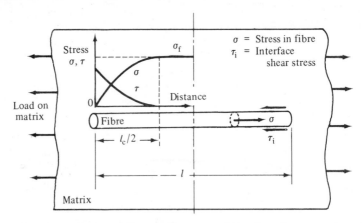

Fig. 1.19 *The distribution of fibre tensile stress and interface shear stress in a short-fibre composite.*

critical length l_c it cannot be stressed to its breaking point by the matrix, and is not therefore used to full advantage. The strength of a composite containing short fibres can be written

$$\sigma_c = \sigma_f V_f(1 - \phi . l_c/l) + \sigma'_m(1 - V_f), \qquad (V_f < V_{min}) \qquad (1.7)$$

where ϕ is a function defining the manner in which stress builds up from the fibre ends. If the ratio l/l_c is as low as 10, 95 per cent of the strength obtainable with continuous fibres can still be obtained with discontinuous ones.

The tensile strength of an aligned composite is also highly directional. It is limited by the mechanism of failure which itself changes with orientation. When the composite is stressed parallel with the fibres, failure begins by the breakage of a fibre at some point of weakness and the load that was carried by the fibre is transferred back into the matrix over a distance $l_c/2$ from each of the broken ends. The local increase in stress resulting from this redistribution will in due course break other fibres in the vicinity, and ultimately the whole composite will fail at that cross section. If the load is applied at 45° to the fibres, however, the resolved shear stress for yielding in the matrix or slip at the fibre/matrix interface will be reached well before the fibre fracture stress, and the composite will fail by shear. When the load is applied at 90° to the fibres it is the tensile strength of the matrix or the normal failure stress of the interface (modified by the plastic constraint factor) that results in composite fracture. A simple maximum stress theory of failure[21] has been shown to work very well for some aluminium/SiO$_2$ composites, and for various laminated materials with alternate layers at $\pm \theta$ to the applied stress, but is usually unsatisfactory for unidirectional composites. Tsai[22] shows that in epoxy/E-glass composites a maximum distortional energy theory fits the experimental results more closely. If σ_c is the composite tensile strength parallel with the fibres (given by eq. (1.6)), σ_t is the transverse tensile strength, and τ_c is the in-plane shear strength, the maximum work theory gives for the orientation-dependent tensile strength, σ_θ:

$$\frac{1}{\sigma_\theta^2} = \frac{\cos^4 \theta}{\sigma_c^2} + \frac{\sin^4 \theta}{\sigma_t^2} + \cos^2 \theta \sin^2 \theta (1/\tau_c^2 - 1/\sigma_c^2) \qquad (1.8)$$

Dimmock and Abrahams,[18] however, show that the maximum stress theory is in somewhat better agreement with the experimentally measured strengths of epoxy/carbon composites (Fig. 1.20). As in the case of elastic

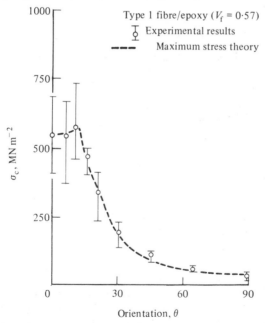

Fig. 1.20 *The orientation dependence of strength of CFRP compared with the predictions of the maximum stress theory, after Dimmock and Abrahams.*[18]

modulus the transverse strength of composites can be significantly improved by laminating cross-plied layers of material.

It is clear from eq. (1.6) that most soft matrices would contribute little to the strength of a composite containing say 50 per cent by volume of high strength carbon fibre. And since the strength of carbon does not change significantly between room temperature and 1000°C the strength of composites should be almost independent of temperature, other factors being equal. In Fig. 1.21 the theoretical strengths of some metal/carbon composites are compared on a strength/density basis with the elevated temperature properties of some common alloys.[23]

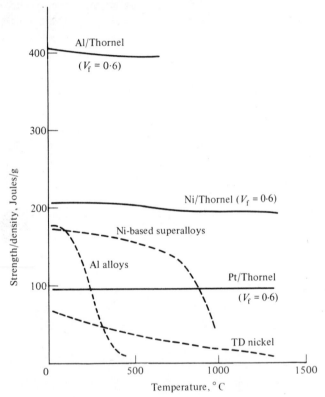

Fig. 1.21 *Predicted variation of the strength-to-density ratio for several carbon fibre reinforced metals, compared with those of some common alloys.*[23]

1.3.3 COMPRESSIVE STRENGTH

When a unidirectional composite is loaded in compression parallel with the fibres, the mode of failure is strongly dependent upon the strength of the fibre/resin bond. If the bond is weak the fibres are debonded from the resin

26

at relatively low stresses, and the compressive strength never reaches the tensile strength. An added complication is that resin cracks can easily occur near the loading point unless lateral spread is restrained. If these limitations are removed, it is still necessary to ensure that the fibres are well aligned in the direction of loading if premature buckling failures are to be avoided. Park has recently shown that compression tests performed in precision-aligned fixtures on type 2 carbon/epoxy composites give results some 30 per cent higher than those obtained on unsupported samples.[24] However, in type 1 fibre composites, in which the resin adheres less readily to the fibres, the increases are as high as 50 to 100 per cent. NOL-ring tests carried out on the same material in tension and compression do in fact give tensile and compressive strengths that are almost identical. Ferran and Harris[25] have shown that carbon fibres embedded in resin are broken into short lengths by combined compressive and shear stresses, and it appears that the degree of lateral support given to the fibre by the matrix could be an important factor in determining compressive strength and the compressive failure of bending beams.

1.3.4 INTERLAMINAR SHEAR STRENGTH

One of the most serious limitations of fibre-reinforced materials is that the shear strength parallel with the fibres can be as low as that of the matrix or the fibre/resin interface, either of which may be an order of magnitude or more lower than the maximum tensile strength of the composite. Consequently the composite may fail at low loads if the stress system is such as to cause a high shear stress on these planes of weakness. A bar loaded in flexure will fail in shear at the neutral plane before it breaks by tensile failure of the outer fibres if the shear strength, τ_{max}, is reached before the outer fibre failure stress, σ_{max}, that is if, in three-point bending, $3P/4wb$ exceeds τ_{max} before $3Pl/2wb^2$ reaches σ_{max}, where P is the load, l is the span, w is the width, and b is the thickness of the beam. For a $0.5\ V_f$ type 1 carbon/epoxy composite the ratio of σ_{max}/τ_{max} is about 50, so that failure will occur by interlaminar shear unless the span-to-depth ratio of the beam is at least 25. The interlaminar shear stress (ILSS) is hard to assess unambiguously, and there is considerable disagreement as to the best method of determining it. It is frequently determined from the failure load of a very short beam in three-point flexure, but it is sometimes difficult to ensure that the failure is uniquely shear.[26] It has been shown that for plates of finite width the ILSS becomes very large at free edges and is small near the centre line instead of being uniform across the plate, as predicted by laminate theory.[27] Thus not only must the geometrical arrangements of test samples and actual components be comparable if satisfactory design is to be achieved, but the rates of loading must also be comparable. For whereas in slow bend tests the high edge stress concentration can be dissipated somewhat by creep in the resin, at high rates

27

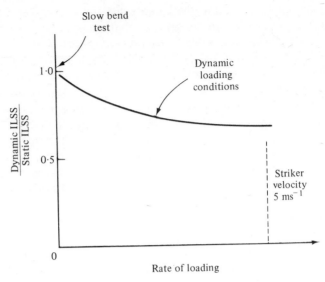

Fig. 1.22 *Effect of rate of loading or impact velocity on the measured ILSS of carbon fibre reinforced plastics (schematic).*

of loading there will be no time for this to occur and the apparent ILSS values measured will be lower, as shown in Fig. 1.22.

The ILSS is governed to a large extent by the strength of the fibre/resin bond since shear failure usually occurs at the interface. The ILSS of CFRP is strongly dependent on the fibre modulus[28] (Fig. 1.23), and this emphasizes a serious disadvantage of the new high-strength fibres. In general the composite shear strength decreases as the degree of graphitization of the carbon

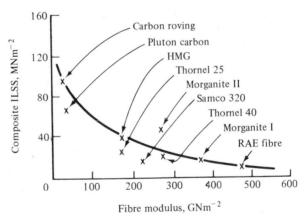

Fig. 1.23 *Variation of composite ILSS with modulus of the reinforcing carbon fibre, after Goan and Prosen.*[28]

28

increases. This is partly due to the fact that the surfaces of carbonized fibres are open and highly porous, whereas after graphitization their surfaces are much smoother. The porosity is caused by loss of material during pyrolysis, and is gradually eliminated during graphitizing treatments as diffusion assists the development of increased perfection in the crystallites. Another factor contributing to the shear strength is the ability of the matrix to wet the fibre surface. The surface energy of the carbon fibre is high because of the presence of active groups. These are fewer the higher the heat treatment temperature and they, too, are removed by graphitization treatments. Because of this problem of low shear strength, a great deal of research has been carried out on methods of improving the fibre/matrix bond. There are two approaches: either the surface can be roughened so as to provide a good mechanical key for the matrix; or some means can be sought of producing a direct chemical bond between fibre and matrix, much as coupling agents do in glass-reinforced plastics. The former method has given the most successful results so far. Methods that have been used are mostly oxidizing treatments, using agents such as nitric acid, sodium hypochlorite, bromine, or simply by heating in air. Commercial treatments such as that used by Morganite Modmor[29] based on a method patented by H. Wells of AERE Harwell are capable of doubling the shear strength to about 40 MN m^{-2} for type 1 fibre composites and to about 80 MN m^{-2} for type 2 composites. Shear strengths as high as 80 MN m^{-2} have also been reported by Prosen who grew SiC whiskers on the surface of carbon fibres.[28] This is rather an expensive procedure, however, and it weakens the fibre somewhat.

As Fig. 1.24 shows[30] the ILSS of Thornel 40/epoxy composites is not strongly dependent on volume fraction of fibre, but whereas it falls with

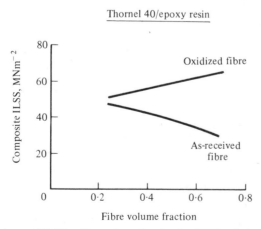

Fig. 1.24 *Dependence of ILSS on fibre volume fraction for CFRP reinforced with untreated and oxidized fibre, after Novak.*[30]

increasing V_f if the fibre is untreated it increases with V_f if the fibre/resin bond is improved by oxidizing the fibre. This simply reflects that in the former case it is the matrix that controls the strength whereas in the latter case the interface shear strength has been increased beyond the matrix shear strength.

1.3.5 TOUGHNESS

Toughness can be defined as the work required to fracture a material, or the fracture energy. In perfectly brittle materials which are elastic to failure the fracture energy is of the same order as the surface energy, usually between 1 and 10 Jm^{-2}. The fracture strength, σ_f, of a brittle solid is very sensitive to the presence of surface imperfections but a high value of fracture energy, γ_F, can help to reduce this notch sensitivity since, according to the Griffith model

$$\sigma_f \approx (E\gamma_F/c)^{\frac{1}{2}} \qquad (1.9)$$

where c is the crack depth. Therefore if any means of absorbing energy during crack propagation can be built into a solid, this will amount to an increased resistance to crack propagation. In metals plastic flow occurs at the tips of sharp cracks which are blunted as a result. The stress concentration is thereby reduced, and the work of plastic deformation may be as high as 10^5 Jm^{-2}. In a composite in which the fibres are strongly bonded to the matrix an advancing crack propagates without hindrance through matrix and fibre and the fracture surface is usually relatively smooth. But if the fibre/matrix bond is weak, lateral tensile stresses in front of the crack tip allow the matrix to become debonded from the fibres ahead of the crack. The fibres are then exposed to a tensile stress over a considerable unsupported length and will fracture at some point of weakness that is not necessarily in the plane of the crack. After the crack tip has passed, the stress at the interface is relaxed and the matrix regains a hold on the broken fibre ends. These must then be pulled out of the matrix before the composite can be separated into two pieces. Two important contributions to the fracture energy are therefore the work of debonding the fibre from the matrix, and the work done against friction in pulling the broken ends out of the matrix. Glass-reinforced epoxy resin has a fracture energy of over 10^5 Jm^{-2}, whereas that of the glass is only about 5 Jm^{-2} and that of the resin only 200 Jm^{-2}. This astonishing increase is probably due in part to fibre pull-out and in part to the elastic energy of fibre/resin debonding. Cottrell[31] has shown that the fracture energy of a composite is proportional to the critical transfer length, l_c, of the fibre in any particular matrix. Since the transfer length is related to the fibre fracture stress, σ_f, and the fibre/matrix interfacial friction stress, τ_i, by the expression

$$l_c/d = \sigma_f/2\tau_i \qquad (1.10)$$

where d is the fibre diameter, the fracture energy is therefore proportional to the product $(\tau_i l_c^2)$ that is, it depends upon both the interfacial friction and

30

upon the distance over which this frictional force acts. In a well-made composite these are not independent, as eq. (1.10) shows. The higher the fibre/matrix bond strength, the shorter is the length of fibre that can be pulled out of the matrix. Therefore if Cottrell's expression is written in the form

$$\gamma_F = \frac{V_f \sigma_f^2 d}{48\tau_i} \tag{1.11}$$

it can be seen that treatment of the fibre surface to improve the interlaminar shear strength will also cause a reduction in toughness. Harris et al.[32] have shown, by modifying the fibre/resin bond in type 1 carbon/polyester composites, that the fracture energy is inversely proportional to the ILSS (Fig. 1.25). However, when the fibre/matrix bond is very weak eq. (1.10) no longer

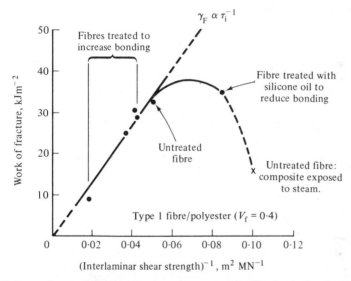

Fig. 1.25 *Dependence of the fracture energy, γ_F, on the fibre/resin bond strength in polyester resin reinforced with type 1 carbon fibres, after Harris et al.[32]*

holds, and as might be expected the toughness of a completely unbonded composite would be low rather than very high as predicted by eq. (1.11). This effect is also shown in Fig. 1.25, where the effect of weakening the fibre/resin bond by coating the fibre in silicone oil and exposing the composite to steam are seen to lower the toughness below that given by eq. (1.11), drastically in the latter case. A designer must therefore be prepared to accept a compromise between toughness and shear strength and must design accordingly. The effect of increasing the fibre volume fraction[33] can be seen from Fig. 1.26: this is in accordance with the predictions of Cottrell's expression. Another important consequence of the pull-out model is that toughness should

31

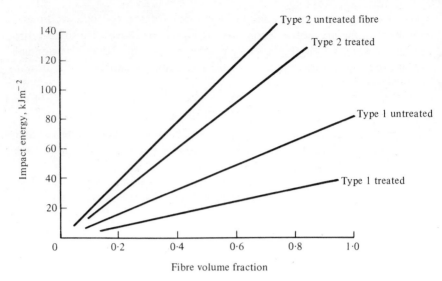

Fig. 1.26 *Variation of fracture energy of CFRP with volume fraction of treated and untreated fibre, after Bader et al.* [33]

increase with increasing fibre diameter, and as discussed in section 1.1 this could have been sufficient incentive for developing carbon fibres of larger diameter than the usual 8 μm. However, the relationship describing the results in Fig. 1.12 is roughly

$$\sigma_f = 4.4 - d/4 \qquad (1.12)$$

where σ is in GN m^{-2} and d is in micrometres, and inserting this into eq. (1.11) gives a polynomial in d for the work of fracture whose numerical value falls with increasing fibre diameter. It does not appear therefore that increases in toughness can be obtained by using thicker carbon fibres. On the other hand, reinforcement with large volume fractions of very fine fibres results in the important phenomenon of crack-suppression in brittle composites. [34] Morley has proposed a way in which the exclusiveness of the requirements for high shear strength and high toughness can be overcome. [35] He suggests that hollow reinforcing elements which are strongly bonded to the matrix be used to provide the required degree of transverse tensile strength and shear strength, and that these contain secondary reinforcing fibres, which after fracture of the outer tube, can pull out over longer distances than in conventional composites by virtue of the controlled degree of friction between the inner and outer elements. He is in fact distinguishing between the fibre/resin interfacial bond strength and the interfacial friction stress under sliding conditions, and it is the latter, determined in this case by the friction between the inner and outer elements, that determines the fracture energy. Early

32

results from composites containing steel wires inside hypodermic tubing have been very successful, but it is difficult to see whether the technique could also work with carbon filaments.

The toughness of composites, like other mechanical properties, is highly anisotropic. Some results of Ellis[36] for type 1 carbon/epoxy composites giving the variation of γ_F, determined by flexural fracture of a notched specimen, show the extent of this anisotropy (Fig. 1.27). For this material the

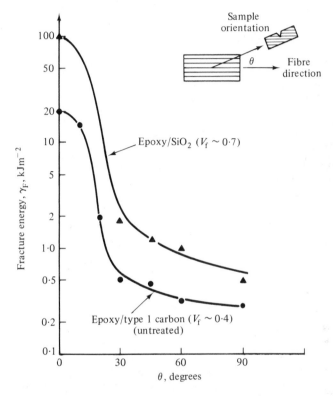

Fig. 1.27 *Orientation dependence of the work of fracture for type 1 carbon/epoxy composites, after Ellis.*[36]

work of fracture for cracking parallel with the fibre direction is some 100 times smaller than that for cracking normal to the fibres. This anisotropy can also be reduced by cross-laminating, when the mode of fracture becomes more complex. Cooper[37] has shown that the toughness of composites reinforced with short fibres is a maximum when the fibre length is equal to the critical transfer length, l_c. Since the strength of a discontinuous fibre composite is only half that of a continuous fibre composite if the fibre length is as short as l_c it is evident that the requirements for high toughness and high

33

tensile strength are also at odds. Again therefore the designer must be certain which is his most important design parameter.

1.3.6 CREEP

The strength of carbon fibres is not strongly dependent on temperature in the expected range of application of composites. And although little creep testing has been carried out on single fibre strands, it seems reasonable to assume that the long-term effect of a load on the fibre would be simply to increase the degree of orientation of the basal planes, and therefore increase the effective stiffness. Normal dislocation-controlled mechanisms of deformation should not contribute significantly to extensions in the fibre direction, although at sufficiently high temperatures diffusion-controlled creep of the Nabarro–Herring kind might contribute small strains. In aligned fibre composites the load will be taken by the fibres, and the weaker matrix will be unable to creep except in so far as the fibres undergo small rearrangements to increase their degree of alignment. Such small strains (of the order of 0·1 per cent) as might accumulate from this source would probably cease after some time.[38] In cross-laminated composites, however, the matrix is more highly stressed and large creep strains will occur with reorientation of the fibres.[39] In the context of creep the problem of the low shear strength again assumes considerable importance if the applied stress system results in a high shear component on planes parallel with the fibres. Here, again, it is the weak matrix or interface which is loaded and the shear component of stress may easily reach a high fraction of the shear failure stress of matrix or interface. Creep will therefore be considerable in such cases. Figure 1.28 gives some indication of the effect of creep (shown here as stress relaxation) in a collection of polyimide/carbon fibre composites shear tested at 150°C.[40] The fraction of the initial applied stress retained after 10 minutes (at constant deflection) is shown as a function of the applied stress. Relaxation is rapid if the applied stress is above half the ILSS.

Creep in short-fibre composites will be determined by the creep behaviour of the matrix under the appropriate degree of elastic constraint due to the presence of the fibres. Environmental effects are particularly troublesome in this respect: the ingress of water, for example, into the interface region will decrease the interfacial friction, which, in turn, will reduce the degree of support given to the matrix by the fibres. The deleterious effect of moisture on the creep properties of glass-reinforced nylon is well known.

1.3.7 FATIGUE

Although there is no information available on the fatigue properties of single carbon fibres, it seems unlikely that their strength will be very cycle dependent. The degree of amelioration of fatigue properties of the composite will again depend upon the relative load supported by the matrix. There is

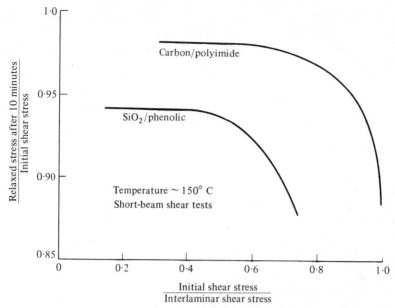

Fig. 1.28 *Stress relaxation in carbon fibre reinforced polyimide. The relaxed stress is plotted as the shear stress after ten minutes under constant deflection (normalized to the initial applied shear stress) versus the initial shear stress (normalized to the instantaneous ILSS).*

already evidence that the strength of aligned, continuous fibre composites is almost independent of number of reversals provided the load is always tensile.[41] As shown in Fig. 1.29 fatigue failures in repeated tension frequently occur only if the applied stress is within the normal scatterband for the static

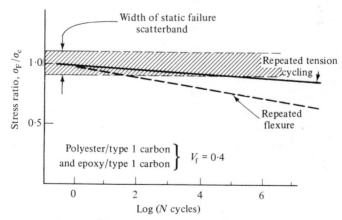

Fig. 1.29 *Fatigue curves for CFRP tested in repeated tension and repeated flexure, after Beaumont and Harris.*[41]

35

tensile strength of the composite, which suggests that long-term failures are a result of a static fatigue effect or simply of gradual failure at statistically distributed flaws and that the important parameter is time under load, rather than number of reversals of load. The long-term fatigue resistance of CFRP at stresses below about 80 per cent of the static strength is therefore likely to be excellent. On the other hand, if the stress system results in a shear component in the matrix or at the interface, fatigue damage, in the form of a reduction in load-bearing ability, occurs at smaller loads and after fewer reversals. This occurs in repeated flexure testing (Fig. 1.29) and in torsion testing.[41]

The same considerations will probably apply to fatigue as to the creep behaviour of composites reinforced with short fibres. The composite fatigue behaviour will be characteristic of that of the matrix under whatever fraction of the total applied load is carried by the matrix after the fibres have taken their share. There is an additional problem, however, if the matrix is brittle or notch sensitive. The ends of short fibres are both points of stress concentration and built-in crack initiators, and their presence will shorten the fatigue life of a composite containing them. In continuous fibre composites, too, after a certain amount of fibre damage has occurred it is likely that fatigue cracks can start at the broken fibre ends.

1.4 Carbon Fibre Composites

Most, if not all, of the illustrations in the previous section are taken from the literature on the mechanical behaviour of carbon fibre reinforced plastics. This field is by far the best developed, largely because it is easier to manufacture CFRP than it is to make carbon fibre reinforced metals and ceramics. In this section we shall consider the progress in the fields outlined in the second section of this chapter, with particular reference to manufacturing methods, quality of materials produced, manufacturing difficulties resulting from incompatibility, and environmental limitations.

1.4.1 CARBON FIBRES IN PLASTIC MATRICES

The incorporation of carbon fibres into plastics results in significant improvements in tensile strength and rigidity. Furthermore, as long as the fibres support most of the load the fatigue resistance is also improved, particularly since because of their thermal conductivity the fibres are able to prevent build-up of heat arising from hysteresis losses, and so reduce thermal degradation which often occurs when unreinforced plastics are subject to cyclic loading. Load cycling in glass reinforced plastics frequently results in the formation of resin crazes which can cause leaks in pressure vessels, or can initiate premature fatigue failure. Crazes occur because of the low modulus of the glass and the high strains that must consequently be used if the full

36

strengthening potential of the reinforcement is to be utilized. CFRP, by contrast, develop high stresses at relatively small strains so that resin crazing does not occur. Design stresses in CFRP can therefore be much higher than in GRP, and this factor must be taken into account when attempting to determine relative cost effectiveness. A degree of anisotropy is impossible to avoid, and must be designed for, but the great advantage that reinforced plastics have over monolithic materials is that structures can be tailored to suit the particular requirements, and local variations of the material's properties can be introduced to cope with the actual stress distribution. This permits more economic use of material. Owing to their brittle nature, the scatter in measured strengths of carbon fibres is quite large. Thus when a composite structure is loaded some of the weaker fibres begin to break at low loads, the stress carried by these fibres being transferred back into the matrix and, subsequently, into nearby unbroken fibres. The final fracturing of a composite is therefore preceded by a continuing, but unobserved process of degradation. This is not usually serious unless the loads applied are close to the composite failure stress. But once loaded a structure will have incurred some degree of damage which weakens it. This fibre damage is often accompanied by small resin cracks and regions of local fibre/resin debonding, and whereas in translucent GRP it can often be detected by optical means, in CFRP it is hidden damage. Non-destructive testing methods that can be applied to large components *in situ* could therefore be of considerable value in determining the degree of damage and residual life or strength. Measurements of the elastic moduli can be made, for example, by determining the speed of sound waves in the material,[42] or the strategic placing of acoustic transducers on the surface of a component can be used to detect and count the numbers and types of event which contribute to the internal damage. It may ultimately be possible to relate plots of accumulated numbers of acoustic emissions as a function of stress, such as that shown in Fig. 1.30,[36] with predetermined measurements of residual strength and fracture toughness.

The limitations of CFRP are frequently associated with resin or interface rather than fibres, and as we have seen, composites usually suffer from low interlaminar shear and fatigue strengths unless fibre surface treatments or a degree of three-dimensional reinforcement can be used. A serious consequence of this is that joining together plates and sections of reinforced material is very troublesome. Bolted joints are suspect, for example, because loads are transferred from one component to another and diffused away from the joint by shear. Joining should be avoided where possible, but combinations of bolting, bonding, and the use of bonded metal shims have been used.

Low compressive strength can often be traced to premature matrix failure, although fibre buckling and fracture have also been found to occur in plain compression tests on columns and in the compressive faces of beams in flexure. This type of damage can lead to poor fatigue properties in samples

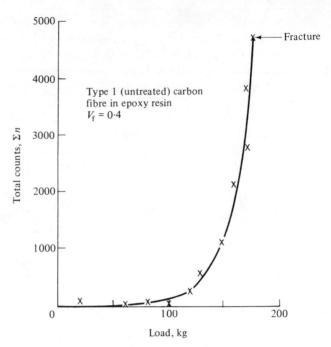

Fig. 1.30 *Acoustic emission from CFRP as a function of load during a flexural test, after Ellis.*[36]

tested in repeated bending by comparison with the same material tested in repeated tension, as Fig. 1.29 shows.

Many common plastics are damaged by exposure to chemically active environments—water, oils, organic solvents, etc.—particularly if they are also stressed. Water can diffuse through plastics and swell their structures so that the fibre/resin bond, or at least the interfacial friction, is reduced or destroyed. Moisture can also be sucked in by capillary attraction along the interface if fibres penetrate the composite surface. The loss of flexural and interlaminar shear strength of composites made of untreated type 1 fibre in epoxy resin exposed to steam and to water is shown in Fig. 1.31.[41] Losses in fracture toughness accompany these strength reductions, although with judicious treatment of the fibre surface this can be prevented. Table 1.2 illustrates the relative efficacy of some surface treatments in this respect. The most satisfactory is an aqueous oxidation treatment followed by coating with a complex organosilane coupling agent such as is used in GRP. This appears to form a mechanically strong, hydrophobic coating which is also firmly bonded to the resin. In contrast, many plain oxidative surface treatments appear to make the fibre surface more hydrophilic which increases the risk of environmental damage.

Tabie 1.2 *Effect of exposure to steam on the work of fracture of type 1 carbon/polyester composites*[32]

Surface treatment of fibre in composite	Percentage of γ_F retained after exposure to steam for one week
Untreated fibre	49
Brominated fibre	56
Fibre etched in boiling HNO$_3$, 2 hours	83
Fibre etched in HNO$_3$ and then coated with epoxy silane (A186)	100

Imperfections introduced into composites during manufacture will affect their mechanical performance. It is true that the art of manufacturing reinforced plastics has reached a sophisticated state of development, and it is unusual in practice to find large amounts of porosity in composites. Nevertheless, poor moulding or laminating practice can have serious consequences, particularly when the matrix itself is called upon to play a role other than that of simply holding the fibres together. Figure 1.32 illustrates the effect of porosity on the fracture toughness of type 1 carbon/epoxy composites normal to and parallel with the fibre direction.[43] While the deterioration is serious in both cases it is worse in the second case where the properties of the resin largely determine the toughness.

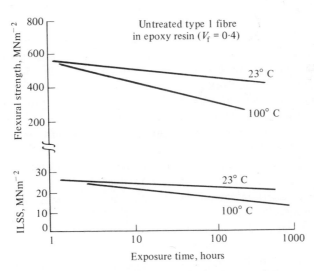

Fig. 1.31 *Loss in flexural and interlaminar shear strength on exposure to moisture at 23°C and 100°C, after Beaumont and Harris.*[41]

39

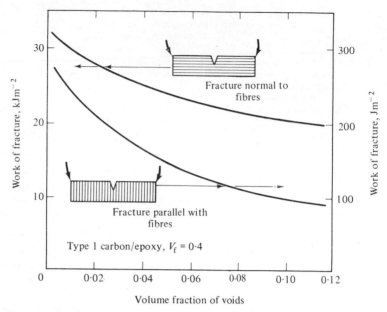

Fig. 1.32 *Effect of void content in resin on the work of fracture of CFRP, after Beaumont.*[43]

One of the consequences of the fact that CFRP are elastic to failure is that the designer cannot make use of plastic deformation to relax stress concentrations at crack tips. Fibre pull-out constitutes an alternative mechanism of energy absorption, and fracture toughness of 300 kJm^{-2} such as is obtained in untreated type 2 carbon/epoxy composites[44] is not far short of acceptable toughnesses for structural materials: it is in fact in excess of that of some aluminium alloys. Unfortunately there is often a large component of shear, and in such cases the fibre pull-out mechanism cannot contribute. Tough composites can thus appear unacceptably brittle if interlaminar shear fracture occurs, particularly under impact conditions when the resin itself is unable to relax edge stresses. A typical example of this effect is perhaps in the bird impact problem which led to the replacement of Hyfil blades in the RB 211 with blades of a titanium alloy. While acceptable strength and rigidity can be designed into such a blade, the laminated nature of its construction can result in interlaminar failure under impact. A possible solution to this difficulty would be the incorporation into the blade of some energy absorption mechanism that can relieve shear stress concentrations. This might be achieved, for example, by modifying the resin with a rubbery phase, a method used to improve the toughness of polystyrene; or by adding a small quantity of metal, in the form of particles or as a wire reinforcement; or by dispersing a small quantity of minute fibres, such as asbestos or some form of

40

whisker, in a random manner to increase the transverse strength and inter-laminar keying. The addition of a metal phase, either as wire or as foil, is also a promising way of tackling the problem of joining in composites. The metal helps to diffuse shear stresses without interlaminar failure and in addition confers a degree of fail-safety.

1.4.2 CARBON FIBRES IN METAL MATRICES

The manufacture of carbon fibre reinforced metals is more difficult than the making of CFRP for two reasons. First, it is difficult to obtain a fully dense solid without high-temperature treatments in the liquid or solid state. Second, few potential matrix metals will wet carbon, so that adhesion is almost always poor. Composites have been made by liquid metal infiltration, by electro-plating followed by hot pressing, by vapour deposition followed by hot pressing, by hot working a mixture of metal powder and chopped carbon fibre, and by electroforming. Many experimental materials made so far have suffered either from excessive damage to the fibres during working operations or from imperfect penetration of the matrix between the fibres. Most workers report that although it is often possible to obtain material with an elastic modulus equal to that predicted by the rule of mixtures, the tensile strength almost always falls short of predicted values. This is usually attributed to the failure of the metal to wet the carbon. Copper and aluminium do not normally wet carbon at temperatures near to their melting point, but if aluminium is in contact with carbon above 1000°C, or if chromium is added to copper above 1150°C both metals will wet the fibre. Unfortunately, the fibre properties are then degraded because of the vigorous attack which occurs. Rolls Royce workers[45] carried out detailed studies of the carbon fibre/nickel system. They found that strengths were usually about half the theoretical value partly because of the poor nickel/carbon bond, partly because of brittleness in the electroplated nickel which prevented load transfer, and partly because fibres were broken during hot pressing. Furthermore, when carbon fibres are in contact with nickel and cobalt for extended periods at high temperatures they are severely weakened by interdiffusion. Braddick *et al.*[46] report that the fatigue properties of carbon/nickel composites are somewhat better than those of the RR 58 aluminium alloy, and although they are slightly more brittle than this alloy they are as tough as boron/aluminium composites. Morris[47] on the other hand reports that the impact energy of 35 vol. per cent carbon/aluminium alloy composites is about $10 \, kJ \, m^{-2}$, compared with 20 times this for unreinforced aluminium and $2000 \, kJ \, m^{-2}$ for the titanium alloy Hylite 45. Workers at the Aerospace Corporation[48] have made liquid metal infiltrated composites using an aluminium/13 per cent silicon alloy as a matrix for Thornel 50 fibres. Perhaps as a result of the high fluidity of this alloy, which is much used in die-casting for this reason, the tensile strength of the composites was about $700 \, MN \, m^{-2}$, equal to the theoretical strength for the material. Pepper et al.[48] point out that they have developed a treatment to promote wetting although they do not give details. It is possible, however, that there is some reaction between carbon and the silicon in the alloy to form silicon carbide. Unexpectedly, the thermal expansion mismatch stresses developed by cycling these composites between $-193°C$ and $+500°C$ caused no damage.

Assuming that the problems of wetting and infiltration can be overcome, there are even more serious difficulties to be tackled in composites that are likely to be exposed to high temperatures or to aqueous environments. First, Braddick et al.[46] have reported that exposure of their nickel/carbon composites to air at 600°C for an hour resulted in serious degradation through oxidation of the carbon at the interface. This is probably the result of diffusion of oxygen through the metal. Second, if aluminium/carbon composites are exposed to water the rate of corrosion of the material in the interface region is accelerated by electrochemical attack. The overall rate of attack is said to be acceptable for many purposes, but the intense localization of the corrosion is a serious disadvantage. One method of coping with these incompatibility problems is to use a coating on the fibre which isolates it from contact with the matrix or acts as a barrier to diffusion of gases dissolved in the matrix. For example, coatings of titanium carbide have been deposited on carbon fibres by vapour deposition.[49] The fibre properties are slightly degraded by the process, but the coating is well bonded and stable, and no further degradation of the fibre occurs after heating for 100 hours in vacuum at 1000°C. It is not yet known whether the titanium carbide could also prevent oxidation of fibres in a composite.

Perhaps the greatest uncertainty at present is whether or not carbon and boron fibre reinforced metals can compete with whisker/metal composites. Apart from manufacturing difficulties, carbon and boron composites exhibit poor mechanical properties in the transverse direction so that as far as the designer is concerned they tend to be heavier and have poorer properties than reinforced resins, and are little better than some of the best unreinforced alloys. Work at ERDE, Waltham Abbey, on the other hand,[9] has resulted in the preparation of whisker-reinforced aluminium alloys, using a pressure die-casting type of infiltration technique, with tensile properties as high as 1000 $MN\,m^{-2}$ in the fibre direction, and because the fibre alignment is not quite perfect these alloys also have remarkable transverse strengths. Some 20 vol. per cent of whiskers having an approximate 'reinforcing strength' of 5 GN m^{-2} are used to achieve these properties. The carbon/whisker competition is seen here at its keenest.

1.4.3 GLASSES AND CERAMICS REINFORCED WITH CARBON FIBRES

Few results have so far been reported for these materials. Workers at Harwell[50] have produced carbon fibre reinforced glass with ten times the strength of plain glass that can be quenched from 600°C without cracking. NPL workers[51] have reinforced silica glass with 50 vol. per cent carbon fibre, and were able to quench into water from 1200°C without cracking, presumably because of the similarity of the expansion coefficients of the components. This composite was found to have many voids at the fibre/matrix interface, and the weaker the fibre/glass bond, the tougher and stronger was the material. The work of fracture was of order 11 kJ m^{-2}. By contrast, the properties of carbon fibre reinforced ceramics described by Bowen[14] are much less satisfactory. For example, although magnesia reinforced with chopped fibres is tougher and more resistant to thermal shock than the unreinforced ceramic, the strength of the composite is many times less. The hot-pressed composite contained many cracks, which is usual in ceramics reinforced with any fibre where there is a thermal expansion mismatch, and although the material was not degraded on heating at 1200°C in an inert atmosphere, heating in air would result in very rapid attack of the fibres by oxygen penetrating the network of cracks. The toughness of magnesia was increased from 10 J m^{-2} to 150 J m^{-2} by reinforcement with 15 vol. per cent of fibres, but this is still low for a structural material. A recent Harwell Materials bulletin (August 1972) summarizes some of the most important developments in carbon fibre reinforced glasses, ceramics, and cements, as follows: (a) Reinforcement of these brittle materials with carbon fibres increases their resistance to catastrophic failure, and improves their thermal and mechanical shock resistance and notch sensitivity. Fracture energies are between 1 and 10 kJ m^{-2}. (b) In inert atmospheres, the strength of reinforced glasses and glass-ceramics is limited only by the matrix viscosity. (c) The fatigue limit of carbon reinforced cement is well above the matrix cracking stress, and the breaking strength of carbon reinforced glass is unchanged after more than 10^4 cycles at 70 per cent of the UTS. (d) 2 vol. per cent of dispersed carbon fibre increases the strength of cement sixfold, and 8 vol. per cent reduces the creep rate of cement paste by a factor of five. The evidence so far militates against reinforcement of ceramics but in favour of reinforcement of glasses which could be the logical successors to the best composites based on thermally stable polymers.

1.4.4 CARBON REINFORCED CARBON

The ideal carbon/carbon system is perhaps based on a vitreous carbon matrix. This is highly resistant to chemical attack, high-temperature oxidation, and erosion by liquid metals, and its inherently low rigidity can be improved by reinforcing with high modulus fibres. In principle this could be achieved by carbonizing composites consisting of a 'non-graphitizing' resin, such as phenol-formaldehyde, in which carbon fibres had already been incorporated by conventional processing methods. Complex parts could be easily manufactured by normal polymer moulding methods and later converted to a high strength material that would otherwise require sophisticated shaping

43

methods. Unfortunately, the high shrinkage of the resin during pyrolysis, combined with the rigidity of the fibres, results in a highly porous, cracked matrix with low strength. Several authors have reported, however, that reimpregnation of the porous material, followed by further carbonization, or the deposition of pyrolytic graphite inside the porous matrix results in materials with the theoretical composite properties. For example McLoughlin[52] describes carbons reinforced with aligned type 1 fibre, the fibres making up some 75–80 per cent of the total weight, which have densities about 1·5, strengths of nearly 1 GN m^{-2} at room temperature, slightly higher strengths at 1500°C (in helium), and fracture energies comparable with those of the precursor epoxy/carbon. Properties of carbon/carbon composites produced by a method operated industrially by Fordath Ltd have been described by Parmee.[53] Although he reports rather poorer properties than those of McLoughlin, he observed that the best flexural strength obtained (about 500 MN m^{-2}) is still 10–15 times that of the best synthetic carbons and graphites currently used in rocket nozzles. However, since the matrix is not the ideal, impermeable, vitreous form of carbon, the oxidation resistance of these materials may not yet be acceptable.

Acknowledgement

I would like to record my gratitude to Dr Alan Baker for several thought-provoking discussions, the seeds of which I have incorporated in this review and to Dr P. W. R. Beaumont and Dr C. D. Ellis for permission to use the results of unpublished work.

References

1 BACON, R., *J. Appl. Phys.*, **31**, 283, 1960.
2 WATT, W., *Proc. Roy. Soc.*, **A319**, 5, 1970.
3 JOHNSON, J. W., MARJORAM, J. R., and ROSE, P. G., *Nature Lond.*, **221**, 357, 1969.
4 KAWAMURA, K., and JENKINS, G. M., *J. Mat. Sci.*, **5**, 262, 1970.
5 HAWTHORNE, H. M., *International Conference on Carbon Fibres, their Composites and Applications*, Plastics Institute, London, 1971 (published 1972).
6 FOURDEUX, A., PERRET, R., and RULAND, W. O., *General Structural Features of Carbon Fibres*, Technical Report 61/70, European Research Associates, s.a., Brussels, 17 August, 1970.
7 JOHNSON, D. J., and TYSON, C. N., *J. Physics, D.*, **2**, 787, 1969.
8 JOHNSON, J. W., *Applied Polymer Symposia*, number 9, 229, Wiley, 1969.
9 PARRATT, N. J., and others, *New Technology*, number 46, May, 1971 (Dept. of Trade and Industry).
10 DE LAMOTTE, E., and PERRY, A. J., *Fibre Science and Technology*, **3**, 157, 1971.
11 MORETON, R., *International Conference on Carbon Fibres, their Composites and Applications*, Plastics Institute, London, 1971 (published 1972).
12 ALLEN, S., COOPER, G. A., and MAYER, R. M., *Carbon Fibres of High Modulus*, IMS Report number 7, National Physical Laboratory, Nov. 1969.
13 BIGGS, W. D., *Contemporary Physics*, **8**, 113, 1967.
14 BOWEN, D. H., *Fibre Science and Technology*, **1**, 85, 1968.
15 CAHN, R. W., and HARRIS, B., *Nature Lond.*, **221**, 132, 1969.

16 HASHIN, Z., and ROSEN, B. W., *Trans. A.S.M.E.*, *J. Applied Mechanics*, **E31**, 223, 1964.
17 TSAI, S. W., ADAMS, D. F., and DONER, D. R., NASA report CR-620, Nov. 1966.
18 DIMMOCK, J., and ABRAHAMS, M., *Composites*, p. 87, December 1969.
19 GOATHAM, J. I., *Proc. Roy. Soc.*, **A319**, 45, 1970.
20 NIELSEN, L. E., and CHEN, P. E., *Journal of Materials*, **3**, 352, 1968.
21 STOWELL, E. Z., and LIU, T. S., *J. Mech. Phys. Solids*, **9**, 242, 1961.
22 TSAI, S. W., *Fundamental Aspects of Fibre Reinforced Plastic Composites*, Schwartz, R. T., and Schwartz, H. S., eds., Interscience, 1968.
23 NIESZ, D. E., Battelle Memorial Institute, Eighth Bimonthly Progress report on contract NOw-65-0615-c, Nov. 1966.
24 PARK, I. K., *International Conference on Carbon Fibres, their Composites and Applications*, Plastics Institute, London, 1971 (published 1972).
25 DE FERRAN, E. M., and HARRIS, B., *J. Composite Materials*, **4**, 62, 1970.
26 MULLIN, J. V., and KNOELL, A. C., *Materials Research and Standards*, **10**, 16, 1970.
27 PIPES, R. B., and PAGANO, N. J., *J. Composite Materials*, **4**, 538, 1970.
28 GOAN, J. C., and PROSEN, S. P., *Interfaces in Composites*, A.S.T.M., **STP 452**, 3, 1969.
29 AITKEN, I. D., RHODES, G., and SPENCER, R. A. P., *Development of a Wet Oxidation Process for the Surface Treatment of Carbon Fibres*, Seventh International Conference on Reinforced Plastics, Brighton, October 1970.
30 NOVAK, R. C., *Composite Materials: Testing and Design*, A.S.T.M., **STP 460**, 540, 1969.
31 COTTRELL, A. H., *Proc. Roy. Soc.*, **A282**, 2, 1964.
32 HARRIS, B., BEAUMONT, P. W. R., and DE FERRAN, E. M., *J. Mat. Sci.*, **6**, 238, 1971.
33 BADER, M. G., BAILEY, J. E., and BELL, I., Dept. of Metallurgy and Materials Technology Report, University of Surrey, Dec. 1970.
34 COOPER, G. A., *The Structure and Mechanical Properties of Composite Materials*, Technical Report 15/71, European Research Associates, s.a., April 1971.
35 MORLEY, J. G., *Proc. Roy. Soc.*, **A319**, 117, 1970.
36 ELLIS, C. D., University of Sussex, D.Phil. Dissertation, 1973.
37 COOPER, G. A., *J. Mat. Sci.*, **5**, 645, 1970.
38 DOBSON, B., Ph.D. Thesis, University of Southampton, 1970.
39 SOLIMAN, F. Y., *Composite Materials: Testing and Design*, A.S.T.M., **STP 460**, 254, 1969.
40 HARRIS, B., Unpublished results of work carried out in the course of a Summer Vacation Consultancy at the National Physical Laboratory, Teddington, 1969.
41 BEAUMONT, P. W. R., and HARRIS, B., *International Conference on Carbon Fibres, their Composites and Applications*, Plastics Institute, London, 1971 (published 1972).
42 REYNOLDS, W. N., *Plastics and Polymers*, 155, April 1969.
43 BEAUMONT, P. W. R., D.Phil. Dissertation, University of Sussex, 1971.
44 SIDEY, G. R., and BRADSHAW, F. J., *International Conference on Carbon Fibres, their Composites and Applications*, Plastics Institute, London, 1971 (published 1972).
45 JACKSON, P. W., and MARJORAM, J. R., *J. Mat. Sci.*, **5**, 9, 1970.
46 BRADDICK, D. M., JACKSON, P. W., and WALKER, P. J., *J. Mat. Sci.*, **6**, 419, 1971.
47 MORRIS, A. W. H., *International Conference on Carbon Fibres, their Composites and Applications*, Plastics Institute, London, 1971 (published 1972).
48 PEPPER, R. T., UPP, J. W., ROSSI, R. C., and KENDALL, E. G., *Metallurgical Transactions*, **2**, 117, 1971.
49 PHILLIPS, K., PERRY, A. J., HOLLOX, G. E., and DE LAMOTTE, E., *J. Mat. Sci.*, **6**, 270, 1971.
50 ROBERTS, F. J., and MORRIS, B., *New Scientist and Science Journal*, **68**, 8 July, 1971.
51 CRIVELLI-VISCONTI, I., and COOPER, G. A., *Nature Lond.*, **221**, 754, 1969.
52 MCLOUGHLIN, J. R., *Nature Lond.*, **227**, 701, 1970.
53 PARMEE, A. C., *International Conference on Carbon Fibres, their Composites and Applications*, Plastics Institute, London, 1971 (published 1972).
54 ANON, *Composites*, **3**, 147, 1972.

2. Polymer Matrix Materials

M. Molyneux

2.1 Introduction

With the realization, some 40 years ago, that excellent properties were practically possible with fibre reinforced composite materials a great deal of effort has been spent on the investigation of both fibres and matrix materials. Unfortunately technology developed faster than the understanding of the principles involved and it is only in the last decade that the importance of choosing the correct matrix has been established. Even so, decisions are often made as a result of experience rather than from knowledge of basic principles. This should not be viewed with too much concern because metallurgists, even with their long history, are still at a loss to explain certain strengthening mechanisms by alloying, yet it is done effectively.

Polymer matrix materials are, as yet, by far the most widely used in carbon fibre composites (CFC). They offer a convenient combination of properties and processability which cannot be equalled by any other matrix material. This is understandable if one considers the alternative materials such as metals, alloys, graphite, glass, ceramics, and cements. They are so restrictive in their properties and processability that they do not yet threaten the monopoly that polymers enjoy.

The word polymer is derived from the Greek, *poly* meaning many, and *meros* meaning parts, i.e., a material of many parts. This is a more descriptive term than the misnomer, plastics. The organic polymers of the thermosetting type are used predominantly in CFC. These become hard, infusible solids when subjected to a cure reaction and will be described fully later.

The matrix carries out the following functions in a composite:

(a) It protects the surface of the individual fibres so that the properties are not affected by abrasion or fibre:fibre contact.
(b) It prevents crack propagation through the strong brittle phase by keeping the fibres separate.
(c) It provides a medium by which load is transferred to the fibre.

There are many other properties of the composite which are dependent upon the matrix and include impact, fatigue, and creep resistance plus

46

environmental resistance to high temperatures, moisture, and corrosive liquids. In this chapter on polymer matrix materials, an appraisal of the polymers currently in use with CFC will be given, plus an indication of the polymers that may possibly be used in the future. In this way it is hoped to provide the engineer with sufficient information to enable him to make a preliminary selection of a polymer matrix for a particular application, and then, forearmed, go and discuss the advantages and disadvantages with the suppliers. This type of knowledge will be essential because the large number of applications for CFC will result in an equally large number of property requirements from the polymer matrix due to the different conditions under which the composite will function.

2.2 Polymer Types

There are six polymer types currently in use, or being developed for use, in CFC and they can be listed as follows:

(a) epoxies
(b) polyesters
(c) phenolics
(d) silicones
(e) polyimides and other high temperature polymers
(f) thermoplastics.

Each type can be modified to provide a wide spectrum of properties. All, with the exception of the silicones, are organic. Before describing them in more detail it is relevant to explain the mechanism of the chemical reactions involved during the manufacture and processing. In the first instance the polymers are synthesized from individual constituents known as monomers and this can be described as the joining together of the many parts as illustrated earlier. The result is a linear long-chain molecule of high molecular weight. At this stage the polymer is combined with the carbon fibre and a further chemical reaction is then induced by the addition of hardeners or catalysts. This is known as the cure and completes the formation of a tightly bound three-dimensional molecular network. The polyimides and thermoplastics do not undergo the curing reaction but they have the desired properties nevertheless. A simple schematic representation of the chemical processes is shown in Fig. 2.1.

To discuss the polymers in more detail it is convenient to separate them into a performance category. Their operating temperature range allows a well defined classification into three groups.

2.2.1 POLYMER MATRIX MATERIALS FOR USE AT AMBIENT TEMPERATURES (<150°C)

(a) *Epoxies*. These polymers have been utilized more than any others in CFC. They possess several important properties which make them unique

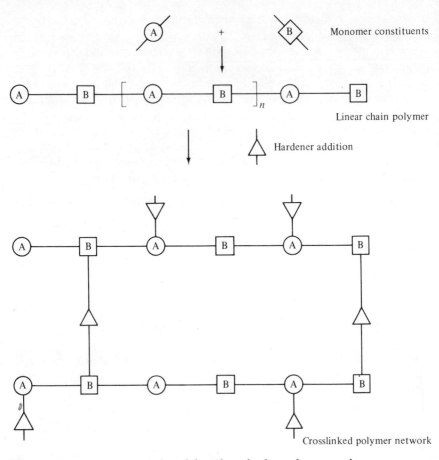

Monomer constituents

Linear chain polymer

Hardener addition

Crosslinked polymer network

Fig. 2.1 *Schematic representation of the polymerization and cure reactions.*

amongst the other polymers. Although certain others may exhibit a similar outstanding property, none has the combination that is available in epoxies. The following list indicates the extent of this versatility.

   (i) Wide choice of forms and curing systems.
  (ii) Curing latitude involving time and temperature.
 (iii) Low shrinkage thus reducing residual stress levels.
 (iv) Mechanical properties superior to other polymers.
  (v) Good adhesive properties to fibre.
 (vi) Good chemical resistance.
 (vii) Dimensional stability is good in stressed condition.
(viii) Thermal stability covers wide range.
 (ix) Excellent barrier to moisture and humid conditions.

The five basic types that are used are:

(i) the diglycidyl ether of bisphenol A (DGEBA) which is often referred to as a standard epoxy
(ii) epoxy novolacs
(iii) epoxy cycloaliphatics
(iv) halogenated epoxies
(v) flexible epoxies.

They are all characterized by the epoxy functional group $-\overset{\overset{\displaystyle O}{\diagup\,\diagdown}}{C-C}-$ as can be seen in Fig. 2.2. Each can be cured into the three-dimensional

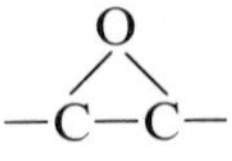

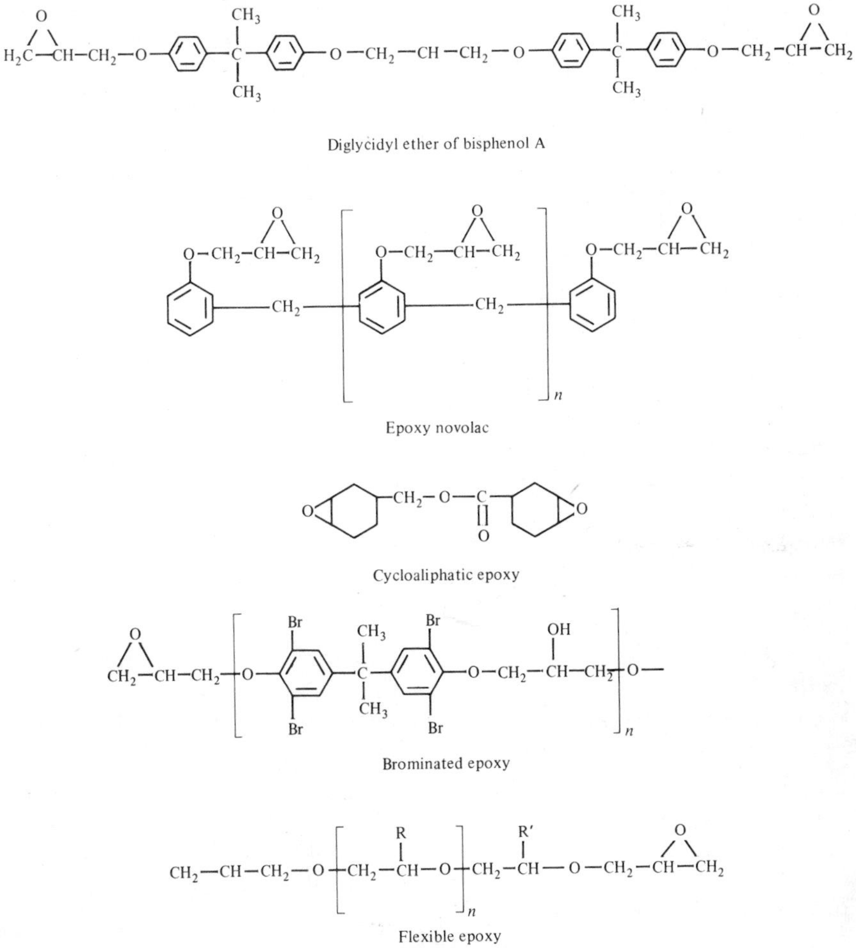

Diglycidyl ether of bisphenol A

Epoxy novolac

Cycloaliphatic epoxy

Brominated epoxy

Flexible epoxy

Fig. 2.2 *Typical chemical construction of epoxy resins.*

49

structure by several hardener:catalysts systems chosen from a range of over 100, the more common being:

   (i) aromatic amines
  (ii) aliphatic amines
 (iii) acid anhydrides
 (iv) polyamides
  (v) phenolic novalacs
 (vi) polysulphides
(vii) complex amine salts.

For the temperature range under consideration DGEBA, brominated, and flexible epoxies are utilized. The DGEBA types are general purpose and can be used in a variety of applications, especially where extreme conditions are not encountered. In the halogenated epoxies bromine is usually linked by a chemical bond onto the main polymer chain although some epoxies are available with chlorine and phosphorus. The main attribute is their flame resistant properties and the demand for this type of polymer matrix in CFC will increase in line with the whole tightening of the regulations governing the fire properties of composites. When bromine content reaches 20 per cent the composites are usually classified as self extinguishing. The operating temperature of these polymers is affected by the addition of bromine and they are not used above 80°C. This is because the principle of flame resistance is dependent upon the release of bromine from the chain, by heat, and it cannot therefore be linked into the chain too strongly.

To obtain a flexible epoxy the standard epoxy is modified by incorporating long-chain glycols or fatty acids into the polymer chain. The epoxy achieves a greater elongation to break because of this and has improved impact properties, but the mechanical properties in general are deteriorated and the temperature capability reduced. Their resistance to solvents and chemicals is also reduced because of the flexible additives which are susceptible to such attack.

(b) *Polyesters*. Thermosetting polyesters were first produced commercially in the USA in 1941. They very quickly established themselves in the glass reinforced plastics industry and their development in this field is well advanced. The main attributes are low cost, fast cure, and good prepregging qualities but the relatively poor mechanical properties has been the main restriction to their acceptance for widespread use with CFC. However, interest in them is growing as it is slowly realized that by changing the basic constituents and adding fillers, flexibilizers, diluents, etc. the required properties can be obtained.

Starting typically from diethylene glycol and maleic anhydride, a polymer chain, as shown in Fig. 2.3 is formed by a condensation reaction. Because of

50

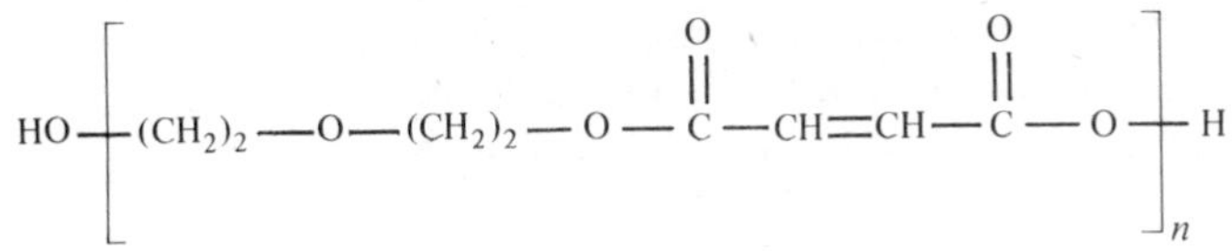

$$HO \left[(CH_2)_2 \text{---} O \text{---} (CH_2)_2 \text{---} O \text{---} \underset{\displaystyle \overset{O}{\|}}{C} \text{---} CH{=}CH \text{---} \underset{\displaystyle \overset{O}{\|}}{C} \text{---} O \right]_n H$$

Fig. 2.3 *Typical chemical construction of a polyester.*

the high viscosity of the polymer upon completion of this reaction, styrene is usually added but this performs a dual role of viscosity controller and cross linking medium. Cure is initiated by catalysts, such as peroxides, and this type of free radical reaction is quite different from the mechanism by which epoxies are cured, and is the reason for its being much faster.

There are several polyesters which are capable of operating at temperatures of 150°C for short periods, but they have the penalty of being excessively brittle. The operating temperatures are usually restricted to the 50 to 100°C range.

(c) *Thermoplastics.* These polymers are quite different from the thermosets, as the name suggests, and they do not undergo a cure reaction to convert them to the solid state necessary for the polymer to carry out the function in a composite. Their physical state is merely governed by temperature. Each thermoplastic has a melting point, dependent upon its chemical construction, above which it can be manipulated into shape. This is the foundation of all the manufacturing processes that have evolved for thermoplastics; the polymer is heated, formed to shape and held in that position until it cools below the melting point so that it retains the shape. All such operations are done mechanically on automated machinery which produces components very quickly. During the past five years glass reinforced plastics have been developed to take advantage of such fast production cycles to produce components with improved mechanical properties. Early work on carbon fibre reinforced thermoplastics was done with these advantages in mind but the results were disappointing due to the severe breakdown of the fibre during the manufacturing process. As the fibres are comparatively brittle and very small in diameter they are broken relatively easily and the pressures and mechanical working of the polymer that is necessary with, for example, injection moulding, can reduce the fibre to dust.

Work is currently in hand to modify the processing machinery to overcome this difficulty because the properties of a thermoplastic CFC will never be good unless a fibre length greater than 1 mm can be maintained.

Certain thermoplastic polymers also have properties which indicate that they would act as good matrix materials for CFC. For example they can be both strong and tough. Nylon, polycarbonate, and polyacetal have been used for early development work specifically for such properties coupled

51

with their higher temperature resistance when compared with other thermo-plastics. Because of the fibre breakdown the properties were not as good as anticipated but sufficient to promote work on better processing methods.

The maximum use temperature will not be greater than 100°C because of creep problems, under load, above this temperature. As the polymer chains are linear and not linked into a three-dimensional structure, thermo-plastics are susceptible to plastic flow at comparatively low temperatures. The exact temperature will of course be dependent upon the particular polymer.

2.2.2 POLYMER MATRIX MATERIALS FOR USE AT INTERMEDIATE TEMPERATURES (150–250°C)

(a) *Epoxies.* The epoxy polymers which extend into this temperature range are the novolacs, the cycloaliphatics, and standard epoxies cured with anhydrides. The novolacs have phenolic groups linked into the chain con-ferring a high functionality and the polymer combines the low shrinkage and freedom from volatiles characteristic of all epoxies, with the high functionality of phenolics. This results in a tightly crosslinked structure when cured which remains stable up to 250°C. This high-temperature resistance is at the expense of toughness and the polymers tend to be very brittle.

The cycloaliphatics have established themselves as the leading matrix material for CFC. They possess a high temperature resistance coupled with very good mechanical properties, including toughness. This is a rare com-bination in a thermosetting polymer and it has been exploited to the full in CFC.

(b) *Phenolics.* These polymers are the oldest and the cheapest thermosets available. They are used widely in the glass reinforced plastics industry for high temperature and electrical applications. Their use in CFC has been restricted because of their poor mechanical properties but several types are in use in high-temperature applications up to 250°C. A serious disadvantage is encountered when moulding phenolic polymers because of the evolution of water, in the form of steam, during the cure cycle. It must be removed otherwise the composite will be badly voided and this reduces the mechanical properties.

(c) *Silicones.* Silicone thermosetting polymers are different from the other polymers previously mentioned inasmuch as they are inorganic due to elemental silicon forming part of the chain. They have similar characteristics to phenolics although certain types can be obtained to operate above 250°C. Their main use is, however, below this at which they retain a very high percentage of ambient temperature properties. The mechanical properties are again poor and this detracts from certain properties in the composite.

52

2.2.3 POLYMER MATRIX MATERIALS FOR USE AT HIGH TEMPERATURES (250–400°C)

Polymers capable of operating above 250°C for long periods first became available in the early 'sixties. This was considered as a major breakthrough and stimulated a great deal of research on different types, which resulted in several new polymers being introduced. A general characteristic of such polymers is an aromatic heterocyclic ring structure which confers high-temperature stability. Such a backbone structure has a very pronounced effect on behaviour because it is very rigid and even though there is no cross-linking it will not yield at higher temperatures. This provides the benefits of crosslinking without the penalties such as extreme brittleness.

The following polymers are the more common high-temperature types; their chemical construction is shown in Fig. 2.4.

(a) polyimides
(b) polybenzimidazoles
(c) polyquinoxalines
(d) polyamide-imides
(e) polybenzthiazoles.

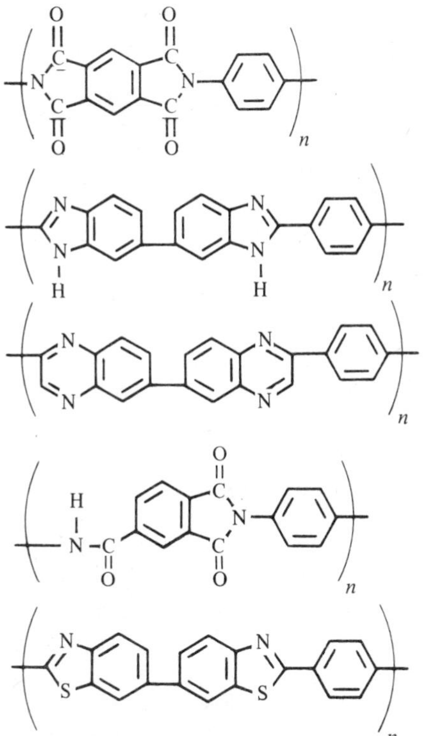

Polyimides

Polybenzimidazoles

Polyquinoxalines

Polyamide-imides

Polybenzthiazoles

Fig. 2.4 ***Typical chemical construction of high temperature polymers.***

The polyimides have been used extensively with CFC but the remainder are still the subject of extensive research and it will be several years before they are used widely.

High polymer melt temperatures and evolution of volatiles during the moulding make them extremely difficult to process. New fabrication techniques are necessary utilizing equipment capable of operating at higher temperatures and pressures. This would mean a further expense added to the already high polymer cost, but the applications for this type of CFC will be specialized and therefore more capable of bearing this cost.

2.3 The Properties Required from a Polymer Matrix Material

The properties of a polymer matrix can be divided into four groups under the general headings, Physical, Processing, Environmental, and Mechanical. These properties govern the suitability of the polymer for use in CFC. By far the most important are the environmental and mechanical properties because of their effect on the structural applications. If suitable properties can be obtained in these areas then a great deal of effort will be put into solving physical and processing difficulties should they be there. On occasions such problems have been insurmountable and polymers, notably some of the high-temperature ones, have been discarded.

2.3.1 PHYSICAL PROPERTIES

The physical nature of a polymer dictates the handlability of the raw materials which are used to make the composite. As the basic starting material is prepreg a great deal of emphasis is put on the ability of the polymer to make good prepreg. Equally as important are the handling characteristics when used directly in the manufacture of a composite without going through the intermediate prepreg stage, filament winding for example.

There are four forms in which thermosetting polymers can be obtained; solid, semi-solid film, liquid, and solution. Because curing agents are invariably added, solid polymers are usually dissolved in a solvent and used in the solution form. This is the most common form as they can be conveniently mixed with hardeners, dispensed quickly and controlled to a fine consistency. Liquid polymers which tend to be viscous are also usually thinned with a solvent to make them easier to handle. Films have certain advantages when making prepreg, but their use is restricted because not all polymers can be made into film.

In making a polymer more adaptable by the addition of solvents certain problems are created. Polyimides for example will only dissolve in strong polar solvents, many of which are toxic, and when heated the vapours given off are extremely dangerous. Certain hardeners are also toxic, especially the amine types which are used with epoxies. They cause skin infections and serious cases can result. Certain people are more susceptible than others and screening tests of production personnel are often necessary.

Many polymers are not particularly stable chemicals and if left for a period of time will undergo polymerization and become hard, even without hardeners. Should these be present, however, the process is accelerated. This phenomenon results in the term 'shelf life' which is used to describe the period of time throughout which the polymer can be processed. The shelf life of solid, liquid, and solution polymers, without hardener, is of the order of months, often years. With prepreg a hardener is present and the cure is advanced slightly to improve handling characteristics. This results in a shelf life limit of approximately three months and indeed with certain polymer systems it can be as short as several weeks at sub-zero temperatures. Storage at low temperatures is recommended for all prepregs to prolong their life and an accurate temperature history of their storage should be kept.

Other factors beside temperature can affect the shelf life of prepreg and moisture is an important one. Amine hardeners used with epoxy polymers are susceptible and absorb moisture from the atmosphere. Certain polyimides are also deteriorated by moisture which causes polymer breakdown by hydrolysis.

2.3.2 PROCESSING PROPERTIES

The polymer matrix dictates the processing techniques when making a CFC. To ensure a quality component the curing schedule of the polymer must not be complicated. To assist processing, the prepreg must have certain characteristics and these are described in the chapter on manufacturing techniques. It is desirable, however, to emphasize the importance of 'tack'. Little can be done to change this property of a polymer which is an inherent characteristic. The addition of tackifiers, such as synthetic rubbers, tends to impair the mechanical properties and they are not used. Temperature is the most convenient method of controlling the tack without altering the other properties. Hot and cold air blowing guns are used to advantage when laying-up prepreg. Another method of partially controlling the tack is to leave a small amount of residual solvent in the prepreg. This is very necessary with solid polymers, otherwise brittle, non-pliable prepreg, which cannot be handled easily, will result. The amount is dependent upon the solvent and the processing method to be employed, because it must all be removed before final cure, otherwise the outcome will be a composite with a high void level due to volatilization of the solvent. Film polymers exhibit a good tack at very low solvent content. This coupled with the useful property of conferring transverse integrity to unidirectional prepreg makes them an attractive choice for certain applications. The tack, however, is achieved by modifying the polymers, often at the expense of mechanical properties and it is a classic case of balancing the fors and againsts with reference to the application.

The cure cycles for the majority of polymers currently in use are long and complicated. At present, with the technology still in the formative stage, this will be tolerated but it is an area which requires a great deal of development. Only the polyesters cure in what can be considered a reasonable time, e.g., 10 to 15 minutes, but the remainder occupy the moulds for an hour or more with the added burden of post cures sometimes in excess of 24 hours. Whilst this situation is not unreasonable for large components the economics for small ones will not allow it, especially as labour and capital equipment costs become a more significant factor than the raw material costs.

The high cure temperatures necessary for the thermally stable polymers require expensive plant on which to carry out the processing. However, the greatest problem with such polymers is the evolution of volatile material during the processing. This is due to the fact that the curing reactions are condensations which result in a by-product and because of the high temperatures involved this is in the gaseous form. If it is not allowed to escape or not removed before completion of cure the same effect occurs as with leaving too much solvent in a prepreg, a voided composite. Unlike solvent, the condensation volatiles are produced continuously as long as the cure reaction progresses. This is a serious drawback and requires accurate moulding techniques and there is only a fine line between producing a good or reject component. Polymer manufacturers are gradually becoming aware of the problem and coming to grips with it. In the next five years it is likely that a new class of thermally stable polymers will emerge because of current research.

With the much utilized epoxy polymers the only inhibitive cure drawback is the length of time required. The minimum time is in the order of one hour and this will have to be reduced. Many other problems that did exist have largely been overcome as evidenced by the number of successful CFCs that have been made.

2.3.3 ENVIRONMENTAL PROPERTIES

The whole point about environmental resistance revolves around the ability of the matrix to carry out the functions described earlier. It is irrelevant what hostile environment is encountered but if it changes the characteristics of the polymer, sufficient information must be compiled to enable an accurate prediction of its performance in use. Because of the long-term nature associated with environmental performance a great deal of information is missing but the work is being done and an overall picture is being put together.

Perhaps the most important environmental property is thermal stability. A polymer matrix can be classified in terms of thermal properties in two ways; thermal resistance and thermal ageing, and although closely associated there is a distinct difference. Thermal resistance describes the performance of a polymer at a specific temperature. In Fig. 2.5 this is explained in graphical form with a cycloaliphatic epoxy and type 1 high modulus carbon fibre. The

56

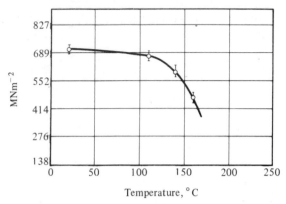

Fig. 2.5 *Flexural strength: temperature, cycloaliphatic epoxy system: ERLA4617/DDM, type 1 high modulus fibre.*

strength falls off slowly until a specific point when it proceeds to drop rapidly. This point is associated with the heat distortion point of the polymer alone and the graphs are used to determine the temperature above which a polymer cannot be used. On the particular graph shown this would be approximately 150°C. Most polymers exhibit this kind of behaviour with respect to temperature as indeed do metals at a higher level.

Thermal ageing relates to the ability of a polymer to withstand prolonged exposure to elevated temperatures. Polymer degradation occurs due to oxidative reactions induced by the temperature and this impairs its efficiency as a matrix material and as a result the composite properties deteriorate. In Fig. 2.6 this is shown graphically.

It is convenient to discuss the flame resistance and smoke emission properties of CFC directly after thermal properties. These again, are purely functions of the matrix and have been neglected in the first few years of development. They will become of importance when stricter regulations, governing the use of composite materials in aircraft structures, are put into operation. Apart from the silicones and some phenolics most polymers will burn readily because of their organic nature, unless specially formulated like the brominated epoxies. Such additions of halogens or other flame-retardent materials to the polymer can confer non-inflammable properties to established specifications. Also low smoke emission polymers can be made but the same situation exists of having to balance one requirement against another.

Chemical and solvent resistance varies widely with polymer types. Epoxies have a good all-round resistance to chemicals and solvents but polyesters are susceptible to chemical attack. It is possible to select a thermosetting polymer for use in a CFC which will be resistant to specific chemical and solvent environments.

57

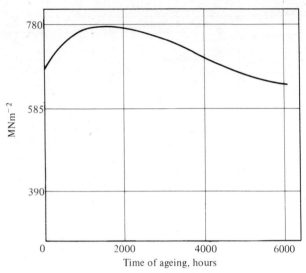

Fig. 2.6 *Flexural strength retention after ageing at 150°C.*

Water, either directly or via atmospheric moisture can have a deleterious effect on CFC because certain polymers are degraded by it. This is especially so with polyesters where breakdown of the polymer chain by hydrolysis can occur. Certain epoxies have shown premature failure when subjected to specific conditions of temperature and humidity and this has caused concern because these polymers were thought supreme. Initial results indicate that the moisture picked up by the polymer is plasticizing it and if this is so the problem can be overcome. Figure 2.7, however, illustrates the fact that not all epoxies are seriously affected.

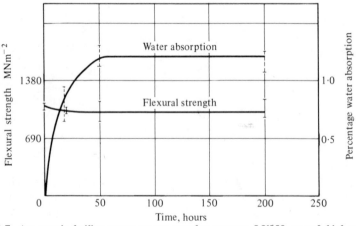

Fig. 2.7 *Age test in boiling water, epoxy novolac system: LY558, type 2 high strength fibre.*

It has been the general practice to base the choice of a polymer matrix for a CFC on the resultant mechanical properties of the composite. Other factors such as the ease of processing and environmental resistance have been regarded as secondary considerations and bearing in mind that the prime function of a CFC is as a structural material it is probably the most suitable criterion on which to base a choice.

The Rule of Mixtures (see p. 20) is the foundation for composite stress calculations and it suggests that the polymer matrix contributes very little to the mechanical properties of the composite. This is not strictly true, but full implications of structural analysis are dealt with elsewhere in the book and it is the intention to discuss here only the properties which are more dependent upon the polymer phase. The following list illustrates the large number of mechanical properties that are required by a stress analyst before he can start thinking in terms of a composite construction.

(a) longitudinal tensile strength and modulus
(b) transverse tensile strength and modulus
(c) longitudinal compression strength and modulus
(d) transverse compression strength and modulus
(e) shear strength and modulus
(f) major and minor Poisson's ratio
(g) longitudinal and transverse strains to failure
(h) thermal conductivity
(i) longitudinal and transverse thermal expansion
(j) fatigue strength
(k) creep resistance
(l) impact strength.

Of these the following are influenced significantly by the polymer matrix.

(a) longitudinal compression strength and modulus
(b) transverse tensile and compression strength
(c) transverse strain to failure
(d) longitudinal shear strength
(e) impact strength
(f) fatigue strength
(g) creep resistance.

Because carbon fibres are highly anisotropic the transverse properties of a unidirectional CFC are very much affected by the polymer properties. This is very evident in the case of transverse compression strength where the actual values of the polymer and composite are always very similar. It is not so noticeable with transverse tensile strength where the effects are probably obscured by the stress-raising effect of the end-on fibres.

The strain, or elongation, to failure of the polymer is a measure of the toughness and ductility and is an important property affecting transverse and crossply properties of the composite. A high strain to failure is desirable and if this can be achieved without sacrificing strength and stiffness so much the better. Due to the large differential in the coefficient of thermal expansion along and across the fibre the residual stresses built up in a CFC with a cross-plied construction, when cooling from the moulding temperature, are very high. The strains are correspondingly high and if the polymer has a low strain to failure, extensive micro-cracks will result. With epoxy novolacs and certain polyimides that are brittle the occurrence of micro-cracks can be heard quite clearly like the snapping of piano wires. Micro-cracks, surprisingly, do not cause an all-round deterioration of the mechanical properties of a composite. The tensile strength, for example, is not significantly affected but it is not desirable to have a composite with flaws present, however small. They can facilitate the absorption of moisture and other potentially adverse materials. Compression properties are reduced as they are very sensitive to the presence of flaws and voids.

Apart from voids, the longitudinal compression strength and modulus of CFC are also dependent upon the modulus of the polymer. Compression failure occurs in two modes but at high fibre volumes the more common is an in-phase buckling of the fibres. If the polymer has a low modulus it will not support the fibres and will allow them to buckle.

The adhesion between the polymer matrix and the carbon fibre influences several composite properties. A unique feature of fibrous composite materials is the very large surface area of contact between the two components. The surface-to-volume ratio, which is dependent only on the fibre dimensions, and therefore large for carbon, illustrates the importance of good bonding. Interlaminar shear strength has been accepted as being indicative of the fibre : polymer bond strength, and a direct correlation between the tensile strength of the polymer and the shear strength of the composite has been established. The higher the tensile strength the higher the shear strength. The adhesive properties of the polymer are also important and because epoxies have very good adhesive qualities they make composites with higher shear strengths.

Associated with fibre polymer adhesion and the interface in general, is the impact property of the CFC. When a brittle material receives a blow it will fracture because cracks open at the point of impact and propagate quickly through the material until complete failure occurs. If the fracture energies of carbon and various polymers are examined separately they are very low, but that of a composite is some thousand times greater. There is one obvious difference and that is the presence of an interface. Because the stress at the tip of a crack is magnified 200 times greater than the initiating stress it propagates quickly. The interface represents a point where the high stress can rupture the bond at right angles to its direction and therefore dissipate the energy. If the bond is very good the crack will penetrate into the strong phase and fracture this until complete failure occurs. As good adhesion is required for high shear strengths a conflict of interests exists and a possible answer is the presence of a controlled bond strength. If a tough polymer is chosen this helps the overall impact strength of the composite as a higher energy is required to initiate a crack but the main answer to the problem of impact resistance is to be found at the interface.

The fatigue strength of CFC does not exhibit a true endurance limit in common with other fibrous composites, but higher results are obtained with the less brittle polymer systems. A certain amount of ductility is necessary to prevent the initiation of damaging microcracks, and to reduce the residual stress after moulding which precipitates fatigue failure. In general the requirements for good fatigue strengths in a CFC are a polymer matrix with a low modulus and a high strain to failure.

Creep experiments conducted from a practical viewpoint to estimate at which point a CFC might fail under service conditions have not been fully completed. However, there are indications that the stiffness of the polymer matrix does play an important part especially when discontinuous fibres are employed in the matrix. The shear modulus of the polymer coupled with the adhesion to the fibre dictates the amount of plastic flow possible. If a thermoplastic polymer is used in a CFC the creep properties will be inferior in comparison to a thermoset because the polymer is not a three-dimensional network and can 'flow' more easily. Again this is highlighted with discontinuous fibres.

From these comments on mechanical properties it is quite obvious that the better the mechanical properties of the polymer, the better the properties one can expect from the composite. This is a generalization and there are contradictions but it is usually correct. This being so, epoxies are by far the better choice and of these the cycloaliphatics stand out. This can readily be seen in table 2.1 which compares a standard epoxy with a cycloaliphatic. In

Table 2.1 *Comparison of unreinforced polymer properties*

		Standard Epoxy DGEBA	Cycloaliphatic Epoxy
Tensile Strength	$MN\,m^{-2}$	68	131
Tensile Modulus	$GN\,m^{-2}$	2·7	4·8
Compressive Strength	$MN\,m^{-2}$	137	193
Compressive Modulus	$GN\,m^{-2}$	4·0	5·2
Strain to Failure	%	3	6

summary the ideal requirements are high strength, high modulus, and a high strain to failure and although these properties may conflict the cycloaliphatic quoted in the table fulfills many of these requirements.

3. Workshop Practices and Processes

M. Molyneux

3.1 Introduction

It is the intention of this chapter on manufacturing procedures for carbon fibre composites (CFC) to convey to the engineer exactly what processes are available so that he has all the necessary information to hand when considering the material for a special application. It is important to be aware of the methods as they exist at present with an indication of the shapes possible, the quantities that can be produced in a given time and with what control of property; plus the associated economics. It is also important, with the rapid advances of technology in the field, to present an indication of possible future manufacturing methods to ensure that the current ones do not restrict the engineer's creativity.

Reinforced plastics have been commercial items for more than 25 years. Glass reinforced plastics especially have established footholds in several major markets. Because carbon fibres are used predominantly to reinforce plastics in the same way as glass, it was a natural beginning to utilize the manufacturing methods evolved for glass reinforced plastics. Whilst this practice can be excused during the first few years of a new material, manufacturing methods that are tuned to its specific processing properties must be developed, otherwise a danger exists in trying to live with, or solve, processing problems that need never be there in the first place. Also, glass reinforced plastics have not reached their ultimate potential, mainly due to an obvious lack of sophisticated and competitive technology for making them and for moulding them into parts. This is not a good recommendation for these methods. Starting with established methods, but never forgetting the need for new concepts, has resulted in not perhaps revolutionary, but certainly evolutionary improvements in manufacturing CFC components.

Two basic requirements are common to CFC moulding; an energy source, usually heat, to convert the plastic material into a formable state and to activate a catalyst which promotes cure; and pressure, which forms and consolidates the material. This chapter describes methods which employ these principles.

Although the majority of methods are not fully automated, a great deal of effort must be spent in this area to progress from the predominantly manual, or with only simple mechanical assistance to manual, operations. If this is not done, CFC will not become competitive with metallic structural materials, regardless of the raw material price. Although the metal industry has had several hundred years to develop manufacturing technology it must not be assumed that CFC requires this sort of apprenticeship. Indeed the rate of growth in scientific and technological knowledge is at such a high compounded percentage to guarantee revolutionary manufacturing methods in less than ten years.

Along with the manufacturing methods, the component quality must reach a high level and be maintained. This may seem an obvious statement of fact, but too often a product suffers in quality because process times have been reduced or the necessary plant is not used because it is thought too expensive. This can prove disastrous in the long-term future of a new material. It is in the early years that component quality must be optimal as unfavourable reputations that are gained can take a long time to erase.

3.2 Raw Materials

Carbon fibre has a less auspicious beginning than raw material for the metal industry where crude ores are dug from the earth's crust. However, just as ores are processed into ingots which may be regarded as a true starting material for component manufacture, so carbon fibres are usually processed into a more amenable starting material by preimpregnation with a resin. The term 'prepreg' is used to describe the combination. Whilst the transformation into ingots is a necessity for metals, it is not essential for carbon fibre to be made into prepreg, but handling of a very delicate fibre is difficult without resin being present. Prepreg may be defined as a pre-engineered-ready-to-mould combination of fibre and resin available in the form of tapes or sheets.

Fibre is available in a variety of lengths from 1 m to 1000 m in filament bundles known as tows. The latter is classified as continuous and the price is as yet almost double that of the short length fibre. When the fibre manufacturers phase out the batch type production in favour of continuous fibre production, the reverse will occur, with short length fibre eventually becoming obsolete. Chopped fibre is available in lengths as short as 2 mm for incorporation into thermoplastics, random moulding compounds, and aligned mats.

Prepreg sheet made from short length carbon fibres is the most common starting material at present. Fibre is laid in parallel rows and impregnated with the desired resin. A consolidating pressure is applied to produce a uniform sheet of known thickness and resin-to-fibre ratio. This unidirectional sheet is the basic building block for subsequent manufacturing operations. The size of sheet is dependent upon the length of fibre used and as this is usually the short length variety the most common sizes are 1 m × 0·33 m with the fibre running parallel to the long length. The thickness of sheet varies between 0·125 mm and 0·250 mm. Thicker sheets present no problem as several standard thicknesses can be plied together. On the other hand, sheets thinner than 0·125 mm are not easy to produce because of the difficulty in separating the 10 000 filament carbon fibre tow. Sheets 0·05 mm, 0·075 mm, and 0·1 mm thick can be made, but the labour involved in manufacture greatly increases the effective price per kilogramme and the uniformity is not good. For thin composites which require multi-axial properties, thin prepreg sheets have to be used to obtain the crossply construction within a given thickness. More effort must be spent over the next few years on improving the manufacture and quality of thinner prepreg. It is now possible for a user, i.e., a moulder, to order a built-up crossply prepreg which can be simply placed directly onto or into his mould. This eliminates cutting and lay-up, which can lead to mistakes and wastage.

Continuous prepreg tape is made from the continuous or long length fibre which is the most expensive at present. As stated previously, this will change and continuous tape will replace sheet; for whatever can be done with sheet can be achieved with tape. The thickness characteristics stated above apply equally to tape, but widths tend to be 75 mm or less because of their use on automatic tape laying machines or in making complex curvature structures over which wide tape will not form.

For the smaller, complex curvature component aligned short staple mat has been developed. It is a prepreg incorporating chopped fibre 2 mm–5 mm long which can be pre-shaped without altering the alignment because the fibres can slide over each other.

Prepregs as starting materials have certain advantages:

(a) Being ready formulated they reduce the chemical knowledge required by a component manufacturer.
(b) No worries about stocking various resins, hardeners, and reinforcements.
(c) Greater design freedom due to the simplicity of cutting irregular shapes.
(d) Waste material is virtually eliminated; offcuts can often be used as random moulding compounds.
(e) Automated mass production techniques can be used.
(f) Quality is reproducible.

There are certain fundamental characteristics necessary in a prepreg so that the user can handle it satisfactorily. Apart from good alignment and

64

fibre uniformity, tack is perhaps the most important. It is difficult to put quantitative units to it, but as a general guide, the release film used as inter-leaving must be removed easily without distorting the fibre and the prepreg should stick firmly to itself but not to the rollers, squeegees, etc., which are used in lay-up. The transverse integrity of a unidirectional prepreg must be good to allow handling during lay-up whether it is by hand or by machine. In certain cases a light woven glass backing is placed on the prepreg to improve transverse integrity. Formability is a property which will generally be good if the tack is correct.

Apart from sheet, preimpregnated or sized fibre tow is used in specific manufacturing processes such as filament winding. A size (5 per cent) of a compatible resin is applied to prevent abrasion and breakage during the process, which may be called wet winding, as the fibre passes through a liquid resin prior to lay-down on the mandrel. With preimpregnated tow the full requirement of resin is present and a bath is not necessary. It is essential that the prepreg tow has the correct degree of tack for this particular process to allow ease of unwinding from the package and prevent resin contamination of guides and rollers.

There are several other forms of starting material in which carbon fibres are incorporated to facilitate the manufacture of certain components. None of these are used to the same extent as prepreg at the present time but their potential is promising.

Carbon fibre filled thermoplastics in the form of granules suitable for injec-tion moulding and extrusion are being developed to take advantage of established manufacturing methods. In these materials 15 to 25 per cent of fibre is added to the thermoplastic, to improve the mechanical, and in some cases wear properties, of components. Problems exist at the moment with fibre breakdown during the process resulting in only marginal increases in property.

Dough moulding compound is a self-descriptive term for a material that has been used for many years with glass as the reinforcing fibre. Components with better strength and stiffness properties in comparison to injection moulded thermoplastics are produced. By substituting glass either wholly or in part with carbon fibre, a combination of high strength and stiffness can be achieved in a versatile moulding compound. A threefold increase in Young's modulus, i.e., from 13 to 40 GN m^{-2} can be realized with small additions of fibre.

The possibility of buying CFC as sheet, tube, and bar stock, off the shelf with accurately specified properties as with metals is not economically feasible at present. However, when the price of carbon fibre falls and reaches a stable level there is no doubt that such a system will begin to operate and the starting materials for CFC components will have taken a major step forward.

3.3 Compression Moulding

Compression or matched die moulding, as it is sometimes called, is a well established technique for reinforced plastics fabrication. It was used on the early 'Bakelites' and the principles basically involve the bringing together of two contoured halves of a mould to impart the shape and cure the resin. The mould halves are heated to effect cure after the material has taken shape. The design of the mould and the plant for combining the two halves under controlled speed and pressure requires careful attention to the moulding properties of the material. With CFC, considerations which radically affect mould and plant design are necessary.

There are no standard shapes of component produced by this method. The main limitation is size, being controlled by the engineering facilities for mould making and the compression presses available. For components larger than 1 metre square other methods are usually found to be more economical because of the high capital outlay on equipment which could only be recouped on high volume production. However, for certain shapes, particularly complicated thick ones, compression moulding is the only suitable method because of the high pressure requirement. Chopped random moulding compounds in particular require pressure in excess of $4 \, MN \, m^{-2}$ to consolidate the material.

3.3.1 PLANT

The main item of plant is a press. More stringent requirements have been placed upon hydraulic press manufacturers in the last ten years and progress from the simple designs of the early rubber presses has been significant. The stimulus for improvement was the development and acceptance of glass reinforced dough moulding (DMC) and sheet moulding compounds (SMC) for volume production. The improvement means that a press can now be bought for moulding CFC which requires only minor modifications to make it completely suitable.

The press should be as versatile and as flexible in its operation as possible, because of the variety of components that it will undoubtedly be required to mould. There are certain requirements that have been established for moulding CFC components.

A frame type of design with down-acting self-contained hydraulics is the most satisfactory system. The guides are very important as they control the alignment accuracy of the two halves of the mould. They should be made out of bronze and be easily replaceable. Platen parallelism needs to be within 0·3 mm per metre and the deflection must not exceed 0·5 mm per linear metre along the diagonal between the frame when loaded over two-thirds of the platen area. The distance between the two platens when they are fully separated, known as the 'daylight', depends entirely on the component.

Usually 500 mm will be sufficient for CFC components. The distance between the two platens when the press is at the end of its stroke is known as the shut height. There are engineering difficulties in having this less than 250 mm but for relatively thin CFC components which are more common than thick, contact of the platens is necessary. An alternative is to put a bolster plate between the bottom platen and the bed of the press to reduce the effective shut height to zero.

The pressure and platen area will depend on the particular type of CFC material being moulded and also the shape of component. However, a versatile combination is 4 MN and 1 metre square to give an effective pressure of $4 \, MN \, m^{-2}$ over the full platen area.

The hydraulics must be smooth acting with negligible delays during speed changes. A rapid advance of 15 m per minute to enable fast contact with the material once placed into the mould is required, ensuring a speedy transfer of heat from both sides of the mould. Once contact has been made it is often desirable to hold in such a position to allow the heat transfer to take place and advance the cure of the resin. This is known as a dwell time and is not usually longer than one minute. The hydraulics necessary for holding in contact position on a down-stroke press are complicated and add expense, but it is an important function. The final pressing speed must be adjustable from 0 to 250 mm per minute. If too fast a closure is used a danger of distorting the aligned prepreg plies exists. After cure is complete the return of the top platen should be rapid, a pullback capacity 20 per cent of rated press capacity should be available although excessive flash, causing the two halves of the mould to jam, is not a feature of CFC moulding.

Heat to cure the resin is supplied via the mould from electrically heated platens. For a large mould, cartridge heaters must be placed at strategic points within it to ensure an even temperature distribution over the surface. The use of thermostats which control the temperature to $\pm 2°C$ are recommended. To contain the heat in the platens and eliminate transfer into the frame, which can cause expansions and buckling, asbestos sheets are used as insulation. If a moulding temperature in excess of 150°C is to be used continually, additional platens which are water cooled should be considered to keep the frame and ram down to almost room temperature.

Automatic control systems can be incorporated to reproduce any cycle of temperature and pressure applications including dwell time. The onus is taken from the operator who has merely to load the CFC lay-up into the mould cavity and initiate the cycle. For long runs this is advantageous as it speeds up the cycle and eliminates human error. However, before an automatic sequence can be put into operation the optimum moulding conditions have to be established for that particular batch of material. Difficulties arise when property variations occur within a batch which results in a poor component because the automatic cycle was not suited to its particular moulding characteristics. Continual examination of the moulding material from the moment it is put into the press is the only way to overcome in-batch variation to which automatic controls are not sensitive.

Equipment that provides a continuous assessment of the material from the moment it is put into the mould is being developed. It is hoped to have a feedback of information direct into the automatic controls so that rather than set a cycle to definite times, the cycle adjusts itself according to the performance of the material in the mould. The principles will be described later in this chapter.

A useful attribute of a platen is the ability for it to be able to move outside of the press frame. Thus when the mould half is bolted to it this is also removed and allows for easy access. It is relatively simple to arrange for the bottom platen to slide forward and because the female part of the mould is usually fastened to this, easier loading, removal, and cleaning of the cavity is achieved. The top platen can also be removed although this is somewhat more difficult. With some types of press used for moulding CFC the top platen has been made to swing down into the vertical position facing the back of the press or alternatively swinging completely over so that it faces upwards. The cycle times for such operations tend to be slow compared to the conventional action but the advantages of access, coupled with the safety aspect of operatives not having to work between the pressure faces, outweigh this.

Before finishing this section on presses it is appropriate to say a little about their location and maintenance. The working area should provide a solid base for the press, preferably concrete at least 200 mm thick. Sufficient room for access to the press for repair and to allow the operator to load and remove the part easily is essential. To avoid point loading of the floor the press may be mounted on a steel plate to distribute the load. Some press makers incorporate this steel plate in the manufacture. A raised concrete flange around the base provides an adequate drip pan for oil leaks and prevents spread. The floor of the area requires sealing to reduce dust which could contaminate the prepreg when brought from the lay-up room.

Maintenance should be preventative as downtime on a press is expensive. A major step can be taken by ensuring that the operators run the press correctly. Posting of a checklist has been found beneficial. A daily inspection by an engineer, taking only minutes, can detect faulty valves, dry bearings, blown heaters, etc., which can be booked for repair at convenient times. Recording all breakdowns provides a useful pattern and can indicate when a press is nearing the end of its useful life. Figure 3.1 illustrates a typical press used to mould CFC.

3.3.2 MOULDS

Good quality moulds are a fundamental requirement for producing CFC components. If adequate moulds are not used tolerances will not be main-

Fig. 3.1 *The extraction of a CFC component from a mould mounted in a compression press.*

tained, surface appearance will be poor, and the properties of the component will be unpredictable and non-uniform.

Metal moulds are the most common for compression moulding when long runs and many components are anticipated. For prototypes, cheaper temporary tools made from wood, plaster, epoxy, and GRP have been used and will be described in a later section.

When specifying a metal mould it is necessary to answer the questions set out in table 3.1.

The design details then require specification and table 3.2 can be formulated.

Meehanite and various steels are preferred to aluminium or kirksite because of their durability. Starting from a casting is cheaper than machining from a billet, but the size of the mould being made will dictate this. Pre-hardened steels (30–35 Rockwell C) allow for easy and accurate machining and critical wear points, such as the shear edge, can be flame hardened (50–60 Rockwell C), when the mould is complete.

The mould surface should be polished after final machining. Chrome plating of 0·05 mm thick will provide a good surface finish on the component, assist release, and also reduce wear, but it is not a necessity.

Ejecting a part from the mould on completion of the cycle has caused problems with CFC. Carbon fibre has a small negative coefficient of expansion

Table 3.1

Specification						
Material and Treatment	Cores					
	Cavities					
	Bolster					
Cavities	Solid		Electro form		Cast	
	Hobbing					
Core Finish	Mirror		Fine Polish		Moulding Min.	
	Textured		Chrome			
Cavity Finish	Mirror		Fine Polish		Moulding Min.	
	Textured		Chrome		Vapour Blast	
Ancillary Equipment	Load Gear		Strip Gear		Drill Jig	
	Clip Tool		Shrink Jig			

along the major axis which means that on cooling, after moulding, the fibre, and hence the component, will expand in this direction. It is obvious that the component could easily jam in a mould cavity under such conditions. Two critical factors affecting this are the pinch-off (sometimes called the shear edge or mould seal) and the system for ejection.

Table 3.2

Ejection	Stripper		Sleeve		None	
	Air		Pin			
Operation	Automatic		Semi-Auto		Manual	
Splits	Mechanical		Hydraulic		Manual	
Inserts	Supply		Spares			
Heating	Holes only		Band Heater		Thermocouple	
	Induction		Cartridge			
Cooling	Simple Runs		Labyrinth		Cascade	
Method of Fixing	Grooves		Slots		Holes	

The first step in effective component ejection occurs during the design itself. Every part must be designed with respect to the intended manufacturing process and ejection is a very critical feature of the process. Obviously, no undercuts can be tolerated as the material will not spring over even the smallest one. Wherever possible, vertical sides should have a 3–5° taper.

Figure 3.2 illustrates a suitable type of pinch-off. It is described as an externally landed, semi-positive system with a vertical flash line providing a

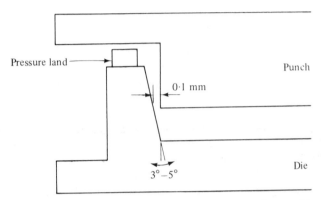

Pressure land

Punch

0·1 mm

3°–5°

Die

Fig. 3.2 *Pinch-off on an externally landed semi-positive compression mould.*

taper on the mould side which prevents the component securing a positive hold and jamming. The flash gap of 0·1 mm applies to aligned continuous and chopped fibre prepreg moulding where only resin is flowed. A similar gap for random moulding compounds can be used but a larger one can be tolerated because fibre is present in the resin flow which makes it less likely to flood out.

3.3.3 PREPARING THE MATERIAL

A lay-up room separate from the moulding area is advisable in order to provide better control of the environment. Temperature is important as it dictates the handleability of the prepreg. Air conditioning units will control the temperature and also the humidity, which can be important when material incorporating a hygroscopic resin is being laid-up. A suitable temperature is 20°C with a relative humidity of 50 per cent.

Dust and other foreign particles adhere very readily to prepreg because of its tacky nature. An extraction system will help, but it is necessary to eliminate as many of the sources as possible at the outset. All types of floor require sealing. The air supply to the conditioner should be filtered unless a good level of purity can be guaranteed. Cleaning rags which may be used should be smooth so that fluff is not produced. Offcuts of prepreg and release film should be placed in polythene bags and sealed. If these straightforward

71

precautions are observed, the possibility of reject components due to contamination will be negligible.

Material arriving at the lay-up room will be in standard size sheets or tape and will require cutting to shape. There are several methods used for cutting which are dependent on the quantity and the size of the shapes required. If there are many, a cutting tool is made and a clicker press used. The tool is simply placed onto the prepreg sheet and fed into the press. The tool is forced through the prepreg leaving a shape with well-defined edges. The technique is very similar to the stamping out of metal blanks. When only a limited number of shapes are required, a template of wood, plastics, or cardboard is used to draw the shape onto the release film. Then scissors or a guillotine are used to cut round the shape so drawn. When the contour is very complex the template can be left in place and a sharp knife used to cut round the edge. The template must be clamped firmly to the cutting bench when this technique is used to prevent distortion of the outer fibres.

The cut shapes are then stacked in correct order prior to lay-up. If the shapes are large each one should be supported individually, especially when being moved, so that fibre distortion does not occur. The release film is left on each individual ply until it is ready for adding to the lay-up. The actual lay-up is usually done by hand, the operatives wearing smooth cloth gloves or non-stick plastic gloves. After individual plies have been added a light pressure is recommended to consolidate them and remove air pockets. This can be done with a roller or squeegee. A warm table can help the consolidation, especially if the prepreg is nearing the end of its shelf life. Such a table is a useful piece of equipment in the lay-up room.

Automatic lay-up has not been developed for compression moulding although a method is established for autoclave moulding. This utilizes narrow tape as the starting material. A similar method for compression moulding will be necessary when high volume production is considered and when labour costs become a more significant proportion of the material costs.

To ensure that the fibre angles of the individual plies are correct as they are laid, appropriate reference lines are drawn on the lay-up table. These are matched to lines drawn on the release film of the prepreg ply. It is helpful to line these up when laying the ply down and then to put a straight edge onto the lay-up afterwards to see if they correspond.

Once the lay-up is complete it should be cocooned in plastic film and placed on a support jig ready for transport to the moulding area.

3.3.4 MOULDING PROCEDURE

The moulding cycle begins when the lay-up is placed into the hot mould cavity. This is a manual operation carried out by the press operator. Automatic loading will be evolved in the future. With aligned CFC prepreg lay-ups, the charge is very similar in shape to that of the mould cavity which

72

means that a positive location is ensured. The two mould halves are brought together immediately until contact is made with the lay-up. Before pressure is applied a period is allowed so that the resin acquires uniform temperature and the resin begins to gel and thicken. As there are many types of resin in CFC and as the gel time is dependent upon the age of the prepreg, gel times vary considerably. For a particular batch of material the gel time must be adjusted to a convenient order, approximately 1–2 minutes, by pre-curing in an oven if it is too long or lowering the mould temperature if it is too short.

For thick section components the resin in the centre of the lay-up will not achieve the same temperature as that nearest to the mould surface over the period of the dwell. No matter how the dwell time is adjusted there will always be a temperature differential causing uneven curing which can result in internal cracks and ply separation. To overcome this partially, the lay-up may be placed into the mould at a lower temperature which is gradually raised to the curing temperature. This detracts from one of the main advantages of compression moulding, fast cycles, and should be borne in mind when considering the economics of the moulding method. An alternative method utilizing high frequency pre-heating is being developed that could very well overcome heating problems with thick section components.

At the end of the dwell time the resin is starting to gel. It is at this point that consolidating pressure is applied, slowly, until the mould is completely closed. The pressure required is dependent upon the material. For long length fibre prepreg $0.6 \, \text{MN m}^{-2}$ is sufficient, aligned mats require $3 \, \text{MN m}^{-2}$ and chopped or random moulding compounds $6 \, \text{MN m}^{-2}$ or greater. The amount of material placed in the mould should be such that 10 per cent to 15 per cent of resin flow by weight is induced.

If the pressure is applied before the resin has started to gel the mould will close very quickly to stops. Gellation and cure will then begin with the resin shrinking away from the mould surface and no pressure being applied to the composite. This will result in a component with a poor appearance and not of good quality. If on the other hand the pressure is applied well into the gellation period the mould will not close and the result will be an oversize, voided component. In the past operator skill was relied upon to ascertain the gel point. Obviously this resulted in many rejects which cannot be afforded especially with CFC.

A technique involving dielectrometry is currently being developed to give the precise moment at which pressure must be applied. A compression mould is fitted with electrodes which are insulated. This is not a simple task with metal moulds as the electrodes must not interfere with the function, and imprints on the component must be minor. The electrodes are connected to a piece of proprietary electronic equipment which measures and amplifies the dielectric constant or the dissipation factor of the resin between them. This is transmitted to a pen recorder which automatically plots the change in property as the resin approaches and passes the gel point. The arrangement can best be described diagrammatically and Fig. 3.3 illustrates the principle. The inflection point of the graph coincides with the gel point and can be picked out easily. The shape of the graph when the component is ready for removal from the mould also gives an indication as to the post cure necessary to achieve optimum properties. This system removes uncertainty from a very critical part of the moulding cycle, i.e., pressure application. It is hoped that

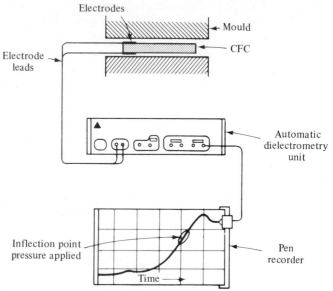

Fig. 3.3 *Arrangement of automatic dielectrometry equipment.*

the dielectrometry unit will eventually form part of the automatic controls of a press so that once a lay-up is placed in the mould the cycle will automatically adjust to the specific requirements of each individual lay-up. The same technique is also being used for autoclave moulding where it is even more beneficial because the component is completely hidden and resin flow cannot be seen. There is one advantage over compression moulding inasmuch as the electrodes can be placed inside the bag lay-up conveniently with no special precautions necessary for insulation.

When a component has been cured in the mould the next step in the cycle is ejection. The significant word is 'cured'; because of the possibility of jamming, discussed previously, it is not advisable to cool the mould before removing the component, and, therefore, the resin must not be flexible when ejected because of the danger of distortion. The stiffness of carbon fibre helps in this respect.

Air ejection has been found suitable for CFC and more especially for thin section components which can easily snap under harsh ejection methods. A poppet valve in the cavity lifts under the pressure of air forced through the passageway leading to it. The air maintains a uniform push on the component once it has been lifted slightly. An ejection sleeve provides another method which gives a uniform push and is to be preferred to pins; however, if these are used they must be applied slowly.

It should not be necessary to cool the component in a jig once it has been ejected. Warpage will occur if the ply construction is not symmetrical and a jig will not prevent this. The post cure, when necessary, is done in air circulating ovens without restraint on the component.

3.4 Vacuum Bag and Autoclave Moulding

Vacuum bag and autoclave moulding is a well established and versatile method that is utilized to manufacture CFC components. The production of aircraft parts by this method has resulted in many improvements in recent years in both plant and technique because of the high quality required.

The method is suitable for prototype components, limited production, production runs of complex components which cannot be done practically by compression moulding, and for large components. The cycle times are longer compared with compression moulding but because the components are usually much larger the weight moulded per hour is often equivalent. Basically this method makes use of a flexible bag or blanket through which an even pressure is applied to a prepreg lay-up on a mould. The pressure is applied by means of vacuum or compressed air, and heat to promote cure of the prepreg is supplied from an oven or autoclave in which the whole assembly is housed. In Figs. 3.4 and 3.5 the layout of an autoclave and oven is shown.

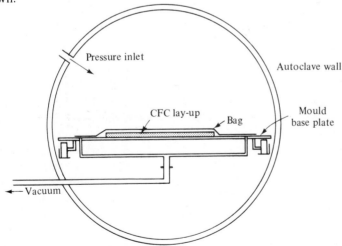

Fig. 3.4 *Schematic layout of an autoclave.*

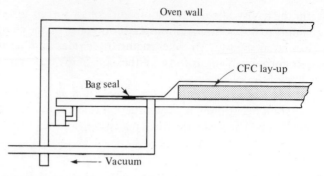

Fig. 3.5 *Schematic layout of a vacuum oven.*

3.4.1 PLANT

The two important items of plant are the oven and the autoclave. An oven is not a sealed vessel and therefore cannot be pressurized. Reliance is put on the vacuum and atmospheric pressure for lay-up consolidation and the bag is kept evacuated throughout the cure cycle.

The oven should be large and convenient dimensions are 2 m wide, 2 m long, and 1 m deep. This allows a wide variety of component shapes to be accommodated. For really large parts, ovens three times the above size are in use. Electric heating elements in the oven are clean and efficient, although for larger sizes, gas heating is more economical. Air circulating fans are necessary to ensure an even temperature distribution. A temperature range up to 300°C should be possible with a control of ± 5°C at any intermediate temperature. The access door can be conveniently arranged to operate in the vertical plane on a counter balance system so that little effort is required in opening and closing.

The dolly, used to support the moulds whilst in the oven, must be robust and supported firmly on guide rails connected to the oven sides. The pipework for the vacuum supply can be installed beneath the dolly plate with just one supply line through the oven side. A high capacity vacuum pump capable of maintaining 65 mm of mercury on all parts of the lay-up is necessary.

An autoclave consists of a large cylindrical metal vessel which can be completely sealed and can therefore be pressurized. Figure 3.6 shows a typical medium size autoclave. Pressures up to $1\cdot3\ \text{MN m}^{-2}$ are common, using air, nitrogen, and carbon dioxide. Heating is achieved by electrical elements situated around the periphery of the vessel with an air circulating fan installed at the back. The door on the front of the vessel must be quick operating and fitted with safety devices. The size of autoclave will depend on the components to be moulded.

Because of the importance of the process it is appropriate to give a more detailed description of the various pieces of equipment in an autoclave.

76

Fig. 3.6 *Placing a bagged part into an autoclave.*

The autoclave case, which is of welded, single wall construction, can be insulated with asbestos cement, fire-brick, or glass fibre. The thickness must be sufficient to keep heat losses to a minimum. The insulation is covered with sheet metal, next to which the electrical elements are situated. These are then covered with a further layer of sheet metal. The number of kilowatts necessary for heating will be dependent on the required temperature and the size of vessel. Temperature control must be maintained to $\pm 5°C$ and this is achieved with a single point controller/verifier via a single thermocouple connection inside the autoclave. More thermocouples at various points in the autoclave are helpful in providing an overall temperature picture when large components are being moulded. If these temperatures are put on a recorder a permanent record of the temperature cycle is obtained.

To maintain an even temperature distribution the air circulating fan at the back of the autoclave is necessary. This is usually driven by an electric motor outside the vessel and the gland which seals the entry for the drive shaft must be carefully designed otherwise it will be a constant source of leakage. An alternative is to completely encapsulate the fan blower unit inside the vessel. This eliminates the need for seals but the cost of protecting the unit from the heat and pressure is reflected in the total price of the autoclave. Around the fan, coils through which cold water can be pumped, are arranged to provide a means of accelerated cooling after cure. This is very necessary to reduce the cycle time which is inherently long to begin with. A cool-down rate from 300°C to 80°C in 45 minutes is of an acceptable order. Other openings in the vessel for thermocouples, gauges, feed pipes, etc., are not difficult to seal as there are no moving parts to contend with and the packing can be clamped tightly.

Because of the large volume within the autoclave and the need to pressurize quickly, a compressor coupled to a large receiver is necessary. Air is the most common pressurizing medium although carbon dioxide and nitrogen are sometimes used mainly because of their fire retardant properties. It is usual to have two pressure controllers in the system, one for low pressure, 0 to 0·3 $MN\,m^{-2}$ and one for high pressure, 0 to 1·3 $MN\,m^{-2}$.

The vacuum supply is standard with one or more connections into the autoclave. It is necessary to apply vacuum pressure before cure and afterwards when the component is cooling down. It is important to have several vacuum gauges, reading from different parts of the bag, to ensure there is no bridging and hence that no air pockets are present.

A further important item is the design of the door. The high pressures necessitate secure locking but this must not be done at the expense of efficiency. The old designs of a multi-bolt-down system are extremely time consuming when opening and closing the door. There are various patented designs now installed on practically all autoclaves which provide a type of cover bolting system operated on either a toothed wheel or sliding pin arrangement. Locking of the door is thus accomplished in one operation by throwing a lever.

3.4.2 MOULDS

A wide range of mould materials can be utilized with vacuum bag autoclave moulding. A decision will depend upon the number of CFC components to be made. If this is more than 100 then it is essential to use steel or aluminium which will stand up to the rigours of repeated processing and reproduce accurate contours. For five or fewer components plaster and wood can be considered and to bridge the gap up to 100 components, glass reinforced plastic or aluminium filled epoxy moulds can be utilized.

The moulding pressure must not exceed 0·15 $MN\,m^{-2}$ with plaster and wood moulds and even then judicious reinforcement may be necessary depending upon the shape. This type of potentially porous mould must be sealed prior to use. When metal moulds are used they should be of a fabricated construction rather than solid to prevent long heat-up rates. The heat-up rates must be adjusted to the mass of material inside the autoclave and large solid moulds act as heat sinks and do not permit rapid heat rise. Information on moulds is given elsewhere in this chapter and applies equally well to the vacuum bag autoclave moulding process.

The same conditions of environment and layout as described for compression moulding apply to the lay-up area for vacuum bag autoclave moulding. Because of the larger components undertaken, larger rooms and lifting equipment are the main extra considerations. There is also a distinct similarity between the two lay-up methods when done manually, one difference with bagged parts being the use of support rollers and gantries to facilitate the lay-up of large areas. However, it is the difficulties associated with manual lay-up that have inspired the design and construction of automated tape laying machines. Unidirectional continuous prepreg carbon fibre tape of 25, 50, or 75 mm width can now be automatically placed accurately onto moulds which are flat or contoured. A lamina layered shell, in which the individual layers are crossplied to any desired orientation that is necessary to sustain eventual working loads, is built up. Tape laying machines are well developed which carry spools of carbon fibre prepreg tape on applicator heads carried and moved by gantry structures which also incorporate traversing, rotating, and tilting mechanisms. Advanced lay-down heads maintain accurate edge matching of the tape by means of sensors, and automatic straight or angular cut-off devices are also installed. Three methods of control are available, (a) manually operated digital control with correction feed-back, (b) semi-automatic and (c) full N/C assisted by an on-line computer.

A first generation lay-down head is depicted in Fig. 3.7 and is typical of a machine used for large flat panel lamination. The head rotates to allow lay-down paths to be oriented with successive layers. When tape is laid onto a compounded curvature there are two geodesic path lengths and soft rubber applicator rollers will not form the tape evenly without creases or splits. The unequal length problem has to be solved by the use of a slack loop in the tape

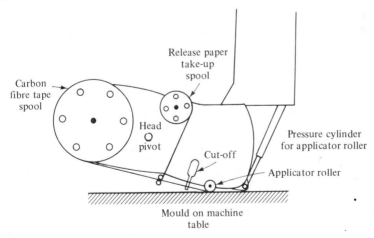

Fig. 3.7 *Simple tape laying head.*

prior to laydown. This complicates head design; however, this type of head is a necessity on certain machines to enable laydown on a mould with complex curvature. Figure 3.8 depicts such a head which incorporates a slack loop device.

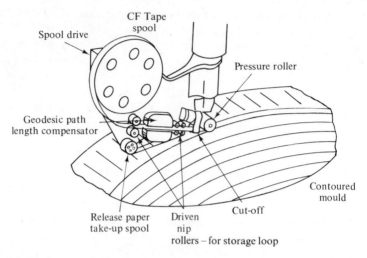

Fig. 3.8 *Modern tape laying head with slack loop device.*

The tape laying machine is an exciting new concept which can promote the use of CFC in the next ten years. It eliminates the time-consuming, expensive hand lay-up technique and provides an accurate, fast method for preparing prepreg ready to mould.

3.4.4 BAGGING

After the prepreg has been laid onto the mould the next step is to assemble the bag and the ancillary materials around it to allow top quality moulding. Several different types of material are built-up in the bag assembly and if care is not taken in this, down to the smallest detail, a reject component can easily result. There are a wide variety of methods used when bagging a component. The choice depends on (a) the shape and thickness of the component, (b) the type of resin, and (c) the pressure and temperature cycle.

A typical bag assembly used for moulding a flat panel is shown in Fig. 3.9. The function of the various materials included is as follows. The release cloth or film placed directly onto the prepreg allows the cured part to be separated easily from the resin flow and the rest of the bagging material. It must be porous to allow the resin flow to pass through it into the layers above. A light glass cloth fabric (50 g m^{-2}) coated with PTFE has been found to work well although for restricted flow, perforated plastic film can be used. Halocarbon

80

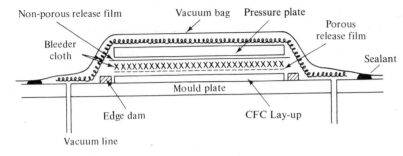

Fig. 3.9 *Typical bag assembly for flat panel moulding.*

film for temperature resistance, with holes 1 mm in diameter punched 20 mm apart is ideal. Layers of bleeder cloth are placed over the release material. The number of plies will be dependent on the thickness of the component being moulded, as this is directly related to the amount of resin that will flow. The bleeder layers serve a twofold purpose of insuring efficient removal, or bleed (hence the name) of air and resin from inside the bag and the absorption of the resin flow. Glass fibre, jute, or paper cloths are used as bleeder material.

A pressure plate as shown in the diagram is often used on flat laminates although this is impractical when shapes are more complicated. Its purpose is to equalize pressure more evenly and produce a smooth top surface. A release film should be placed between the bleeder cloth and the pressure plate to enable removal if the resin flow strikes through all the bleeder layers. A further single bleeder layer is recommended over the pressure plate to eliminate air pockets which could result in uneven pressure application.

Edge dams of compressible cork neoprene composition are placed round the edge of the lay-up to prevent excessive sideways movement and tapering. On occasions the uppermost release film is sealed to the edge dams with masking tape.

Finally the vacuum bag is placed over the assembly and cut so that it extends beyond the bleeder layers which themselves extend 100 to 200 mm from the edge of the component. The bag must be made of a flexible material and the following films are used: polyester, polyvinylchloride, silicone rubber, neoprene rubber, glass coated neoprene rubber, and nylon. The latter at a thickness of 0·03 mm is adaptable to many shapes. Various sealing devices and materials can be used. Clamping over rubber gaskets tends to be difficult with more complicated shapes and the tacky strip sealants of zinc chromate paste or modified rubber are quick to apply and are efficient. An important feature of the sealant is that it must remain stable under the elevated temperatures and pressures experienced during the cure cycle. Should the bag move due to sealant instability the whole part can be ruined especially if this occurs whilst the resin is still fluid.

When the bag has been sealed to the mould the vacuum lines are connected to the base and a vacuum applied slowly. As the bag is drawn onto the part the wrinkles and air pockets are smoothed out, using a plastic rubbing tool or a roller, into the bleeder layers at the edge. When full vacuum is reached in the bag a completely smooth surface should exist.

Several checks for leaks in the bag or at the seal must be made. This is done by shutting off the vacuum supply and noting the gauges to see if there is a drop in the reading. It is almost impossible to obtain a complete seal; 20 mm of mercury per minute can be tolerated. If the leak is greater than this it is most likely to be at the sealing joint.

The bag assembly is ready, at this stage, for loading into the oven or autoclave. The cure cycle for the resin system being used will have been optimized previously and automatic controls, if fitted, must be set accordingly. A typical cure sequence for a CFC with epoxy resin is shown in table 3.3.

Table 3.3

Pressure $KN\,m^{-2}$	Temperature °C	Rate of Increase °C/min
Vacuum	Ambient to 130	2
Vacuum	130	Hold for 30 min
400	130 to 170	2
400	170	Hold for 2 h
400	Cool below 100	

There are several variables that have to be accounted for when optimizing the cure sequence. Prepreg gel time and the temperature heat-up rate are of prime importance with prepreg resin content and component thickness requiring secondary consideration.

The prepreg gel time dictates the point at which pressure must be applied. If it is applied too early and the resin has a very low melt viscosity, excessive resin flow will occur. This can be serious as mechanical properties deteriorate very rapidly above a certain fibre content, albeit high. The gel time on the same batch of prepreg can be determined before cure although this can only relate approximately to the prepreg being moulded because of the possibility of a different heat build-up within the bag assembly. This is an excellent reason for utilizing automatic dielectrometry to indicate the exact point at which pressure can be applied and this is why the method is being developed rapidly as indicated previously. Figure 3.10 shows the use of automatic dielectrometry with an autoclave.

When the laminate thickness of a component is greater than 12 mm, special extended heat-up rates are necessary to prevent cure exotherm causing

Fig. 3.10 *Automatic dielectrometry equipment in use with an autoclave.*

internal cracking of the composite. When thicknesses are greater than 25 mm, cure cycles can be longer than 24 hours.

The magnitude of cure pressure has been found to be non-critical within certain limits. Providing equivalent resin flow is obtained there will be very little difference in the properties of a composite when moulded at 0·3 MN m^{-2} or 0·6 MN m^{-2}. This rule does not apply with pressures of 0·15 MN m^{-2} and below and higher void contents must be expected when only vacuum pressure in an oven is used. On completion of cure the component is kept under pressure, either in the autoclave or under vacuum, until it is at room temperature. It is then removed from the bag assembly and post cured to achieve optimum properties if the resin requires this.

Detailed cure methods for autoclave and oven methods are given below.

Autoclave
(a) Apply vacuum pressure and place the bagged parts into the vessel on the dolly, transferring vacuum lines if necessary.
(b) Seal the autoclave and switch on the air circulating fan.
(c) Switch on the heaters and set the controls for the first temperature step.
(d) When the vessel is pressurized, according to the predetermined cycle, above atmospheric, vent all the vacuum lines to atmosphere.

(e) Once the vacuum lines have been vented frequent inspection for the leakage of internal pressure through them should be made. If it exceeds 20 mm mercury it is an indication that the bag has been displaced.

(f) When the cure cycle is complete, stop the pressure supply and open the discharge valves slowly.

(g) When the pressure nears atmospheric close the vacuum lines and apply vacuum pressure to the component.

(h) Allow the component to cool to room temperature or below 100°C before turning off the vacuum and removing from the bag assembly.

Oven

(a) Apply vacuum pressure and place the bagged parts into the oven on the dolly, transferring vacuum lines if necessary.

(b) Switch on the heating and air circulating systems.

(c) Maintain vacuum pressure and check regularly.

(d) Proceed through the specified temperature schedule until cure is complete.

(e) When the cure cycle is complete shut-off heaters and allow to cool with full vacuum maintained.

(f) When the component has been cooled to room temperature or below 100°C turn off the vacuum and remove from the bag assembly.

On the completion of either of the above moulding operations the care necessary in removing the CFC component from the bag assembly cannot be emphasized too much. With certain parts wall thicknesses are very fragile and the numerous items in the bag assembly tend to become an integral unit. If care is not taken in separation the fragile parts can become damaged.

3.5 Filament Winding

The filament winding process is simple in its conception and consists of winding continuous carbon fibre tow over a mandrel in a machine-controlled operation. Several layers of fibre are put on the mandrel in a variety of patterns depending upon the intended application. Accurate placement of the fibre under controlled tension is an attractive advantage of the process. The carbon fibre tow can be in an unimpregnated form although a light sizing is recommended to prevent abrasion when passing over guide and tension rollers. Alternatively, impregnated tow can be utilized which means that the impregnation tank, through which the tow would have to pass before lay-down, can be omitted. The process is completed by curing the resin at elevated temperature, usually without the application of pressure. The mandrel is removed after completion of cure.

Hollow vessels of a spherical or cylindrical nature are the typical shapes that are produced by the process. Examples include rocket motor cases, nose cones, pressure vessels, rocket missile bodies, and various weaponry barrels on the military side and storage tanks, pipes, bottles, and electrical fittings on the commercial side.

3.5.1 PLANT

There are two basic types of filament winding machine, namely polar and helical winders. With the former the mandrel remains stationary whilst an arm rotates longitudinally round its axis. The arm deposits the fibre onto the mandrel which itself is rotated the exact width of the fibre each revolution of the arm. Several layers of fibre can be put onto the mandrel and it is distinct from the helical winding method in that there are no fibre crossovers. Fibre angles are low and are used in conjunction with hoop windings which a polar winder is capable of doing.

With helical winding, the mandrel revolves continuously whilst a carriage, which dispenses the fibre, traverses back and forth at a speed matched to the mandrel speed to give the required helix angle. The fibre crosses at several points on the mandrel and one complete layer consists of two balanced plies. The control of both types of machine can be by mechanical or numerical methods. With the former, speeds are regulated by gear trains, chains, or feed screws. The system is cheaper but not as versatile as numerical tape control.

Of the two types helical winding is the more utilized because of its inherent flexibility even though the fibre crossover is an accepted stress raiser. By careful selection of winding patterns this can be reduced considerably. Typical winding patterns are shown in Fig. 3.11.

3.5.2 MANDRELS

The mandrel plays an important part in the manufacture of a good-quality CFC component. There are several kinds of material that can be used, each suitable for a particular type of component. The following list indicates the wide choice available depending upon the shape and size of component to be wound:

(a) solid metal, steel or aluminium
(b) water-soluble salts
(c) water-soluble plasters
(d) low melting point alloys
(e) collapsible metal constructions
(f) plastic foams
(g) inflatable.

Solid mandrels can only be used for open-ended components so that they can be extracted. They are best suited to pipes and tubes. The torque transmission tube shown in Fig. 3.12 is typical of the type of component that can be wound on steel or aluminium mandrels. Soluble salts or sand/polyvinyl-alcohol mixtures are extremely versatile and can be used for components up

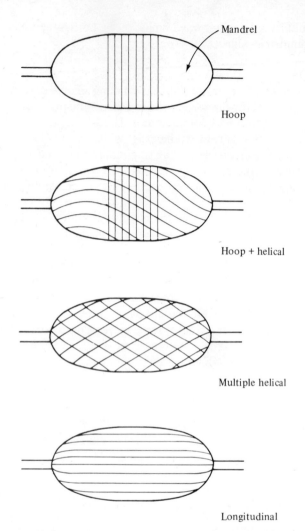

Hoop

Hoop + helical

Multiple helical

Longitudinal

Fig. 3.11 *Filament winding patterns.*

to 500 mm in diameter. For the larger diameters such mandrels should be reduced in weight by the addition of glass reinforced plastics tubing or rigid foam. Soluble plasters are rather weak and their main use is in the facing of metal mandrels. The inability to remove all of the plaster by simple washing procedures is a disadvantage. Low melting point alloys are restricted to small components not usually greater than 250 mm in diameter. They are expensive and their low stiffness results in distortion under moderate loads. Collapsible metal mandrels are only suitable for diameters of 750 mm or greater because of the space necessary for such mechanisms as knuckle or toggle supports

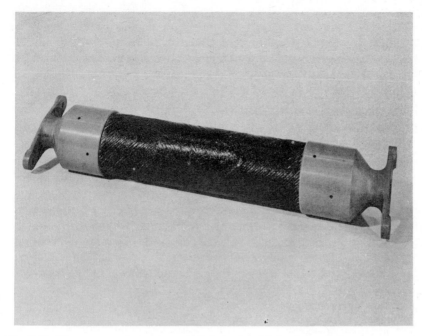

Fig. 3.12 *Filament wound torque transmission tube with bonded end pieces.*

for the mandrel shell. Foams are not recommended because of the poor mechanical properties and the restrictions they impose on elevated temperature curing. Inflatable types are useful for diameters above 125 mm. Inside support is necessary to eliminate the distortion on winding.

All the above types of mandrel can be adapted to the wrapping technique for the manufacture of tubes, described elsewhere in this section, bearing in mind that lighter loads are involved.

3.5.3 WINDING PROCEDURE

Before the winding operation can begin there are several decisions that must be made. Assuming the filament winder and the mandrel are available, the type and form of the fibre has to be specified. If polar winding is being considered then a prepreg tow is necessary to prevent slippage. If unimpregnated tow is preferred a light size is recommended to prevent fibre damage. The choice of resin system will be dependent upon the end-use application of the component but there is a wide choice available either in the impregnated or unimpregnated form. A package of spirally wound impregnated fibre should be supplied with a 10 mm wide silicone paper or alternative release film as a backing to prevent snagging during unwind. The brittle nature of the fibres can result in breakage very easily if a snag occurs. The tack of impregnated tow is of great importance. If the release backing is removed immediately the tow leaves the package it must have the minimum tack so that there is no sticking to the lay-down and intermediate rollers. However, there must be sufficient tack to enable the tow to adhere at steep angles, on the mandrel. Warm air blowers are a useful piece of equipment if there is insufficient tack.

Winding patterns can be resolved into three categories.

(a) low angle polar (longitudinal) with hoop
(b) multi-circuit helical
(c) low angle multi-circuit helical with hoop.

The choice of pattern or combinations thereof will depend upon the component and its application. To maintain dimensional control a number of layers can be wound onto the mandrel and cured, followed by a grinding operation to a predetermined wall thickness. Apart from the fact that fibres are severed, there is also the introduction of an interlaminar discontinuity, thus lowering the shear strength. This may or may not be important, but step winding followed by grinding should be avoided if possible. The final operation in the process is the curing of the resin and removal of the mandrel.

3.6 Wrapping Techniques

There are often occasions when it is necessary to manufacture tubular components by other methods than filament winding. For example when simple lay-up patterns, fast production speeds and low cost are prerequisites it is not economical to filament wind. In such cases semi-automatic prepreg wrapping has been used successfully. Tubes are an obvious component for this method and, indeed, glass reinforced plastic tubes made by this technique have been available for many years. Figure 3.13 shows a racing bicycle made from CFC tubes which were manufactured by a wrapping technique.

Just as with filament winding, mandrels are used for the same purpose and the discussion in the previous section applies equally well to this technique. Prepreg sheet or tape is cut to shape depending upon the type of wrap required. Mechanical assistance is used to consolidate the prepreg on the mandrel when wrapping and pressure can be applied during the cure by shrink film, vacuum bag, or split mould. These processes have resulted in the development of special manufacturing equipment which will be described.

3.6.1 PLANT

A common wrapping/consolidating machine consists of two platens, one of which moves horizontally with respect to the other. The length of traverse and also the width and length of the platens will be determined by the size of tube to be wrapped. Both platens are capable of being heated and coming together under a controlled pressure, the combination of heat and pressure being adjusted to give the optimum prepreg consolidation without destroying

Fig. 3.13 *Racing bicycle made of CFC tubes utilizing various ply orientations.*

the fibre orientation. Wrapping techniques of different principles from the one described are depicted in Fig. 3.14.

Tubular components have been made successfully by hand wrapping onto the mandrel, using a bench and a hand manipulated board to consolidate the prepreg. If there is insufficient tack of the prepreg, poor ply adhesion will occur and a heated table can be used with effect.

Shrink film, e.g., cellophane, is often used as the pressure source during cure. A film wrapper should be used so that a controlled and even tension is ensured during its application. If one is not used there is a strong possibility of uneven wall thickness in the finished tape. Certain skilled operatives can apply shrink film quickly and with even tension, but variability is more likely to occur than with a machine. When the component has been cured the shrink film has to be removed. If the matrix is epoxy resin this can prove difficult

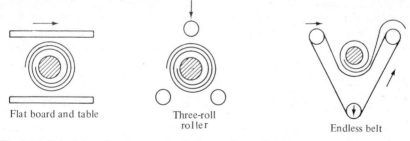

Flat board and table   Three-roll roller   Endless belt

Fig. 3.14 *Principles of various mechanical wrapping techniques.*

because of its adhesive qualities. Slitting machines are used to shred the film without marking the tube and the film can then be brushed off easily. With cellophane, a soaking in hot water is necessary before the slitting operation.

The removal of a mandrel is of course dependent upon the material of which it is made. With metal mandrels which have parallel sections, a mechanical extractor is a necessity. Large frictional forces have to be overcome before the mandrel can be extracted. They are the result of the composite shrinking onto the mandrel when cure takes place. The negative coefficient of expansion of CFC along the major axis has been useful in this respect. It enables a crossply to be made which reduces the transverse shrinkage and hence the frictional forces.

On thin wall tubes, i.e., less than 0·5 mm thickness, with a unidirectional fibre orientation, the compressive strength has been found to be lower than the frictional force which has to be overcome on extraction of the mandrel and, on trying, the tube has simply failed in compression. However, with crossply constructions, tubes with such wall thicknesses can be pulled off a mandrel easily because of reduced shrinkage.

Figures 3.15, 3.16, and 3.17 illustrate the machines described previously. It is pertinent to discuss several points in relation to mandrel extractors.

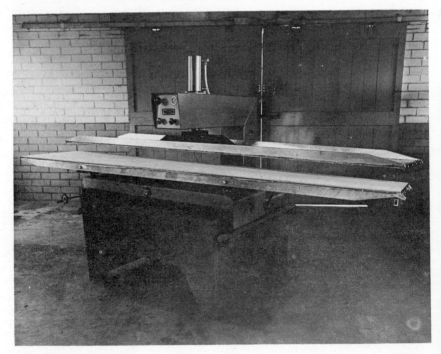

Fig. 3.15 *Mechanical wrapping/consolidating machine (by courtesy of Century Design La Mesa, California).*

90

Fig. 3.16 *Machine for the application of shrink films (by courtesy of Century Design La Mesa, California).*

Variable speed motors as the drive source are simple and provide a smooth take-up which avoids putting shock loads onto tubes. Less expensive drives can be used and include electric motors through a gear box and clutch, and hydraulic systems are also quite successful. For short tubular or tapered

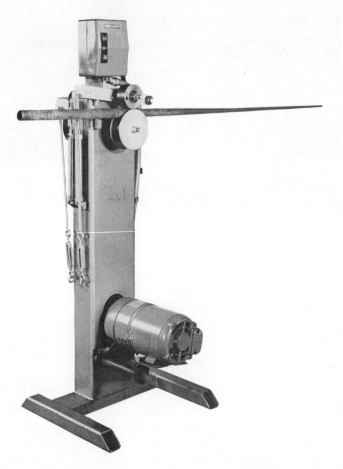

Fig. 3.17 *Shrink film slitting machine (by courtesy of Century Design La Mesa California).*

components, where the mandrel need only be moved a short distance, hand-operated screw type extractors work admirably.

The use of centreless grinding machines to bring the outside dimensions of tubular components to within tolerances is not acceptable. Apart from expensive waste, the fibre damage to a carefully designed crossply construction degrades the properties. Other methods of controlling dimensions and appearance are necessary.

3.6.2 MANUFACTURING PROCEDURES

The surface finish of mandrels that require mechanical extraction must be smooth. A good ground surface rather than polished will suffice but scores or burrs arising through careless handling will create problems on extraction, and cannot be tolerated. Before wrapping on prepreg the application of a release agent is necessary. Silicone wax baked on is very satisfactory for CFC. Other types including PTFE, carnauba wax, and silicone emulsions can be used but vary in performance depending on the resin system.

Films such as cellophane have also been used successfully and for a particularly difficult release a thermoplastic film such as polyethylene or polypropylene has certain benefits. After cure, the tube is extracted hot whilst the thermoplastic is soft with a very low shear strength, thus the tube slides off quite easily. After wrapping, either by hand or machine, a piece of waste prepreg, approximately 10 mm wide, is wound round the mandrel drawoff side of the tube to thicken it up. This is more essential on thin walled and unidirectional tubes as it prevents crushing and splitting of the end on extraction.

A spiral-wrap shrink film is applied on the machine. This type of wrap, although effective, leaves a 0·03 mm spiral ridge on the finished tube which for some applications is not acceptable. Tubular films which shrink biaxially are used when the shape allows and when a completely smooth surface is required. Unfortunately, this type tends to be relatively expensive and is doubtful from an economical standpoint for long production runs.

A method of applying pressure and also of maintaining outside dimensions involves the use of a split mould made out of steel, alloy, or GRP. The lay-up on the mandrel is placed in one half, the two halves clamped together and the assembly placed in an oven to cure the prepreg. Figure 3.18 illustrates the

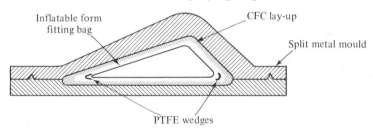

Fig. 3.18 *Split mould and inflatable bag for moulding tubular components.*

principle. The resultant tube will have a flash line where the two halves of the mould meet, but if the correct amount of prepreg is put onto the mandrel this will be very fine. Moulds of any kind are expensive and if a number of tubes have to be made by this method the cycle time has to be short or several moulds will be required. The key to success with the method is to have the correct amount of prepreg on the mandrel. If there is too little there will be no consolidation; if there is too much the excess will be forced into the split line and the mould will not close.

To overcome this critical feature of the mould, expandable mandrels can be adopted and very good tubular components can be produced. Inflatable bags are better in comparison with expanding metal mandrels because of the complicated nature and cost of the latter. Cloth reinforced silicone or neoprene, nitrile, and fluorocarbon rubber expandable bags are the most common. The prepreg can be laid up on a separate mandrel and the bag placed inside after it has been removed or alternatively the bag can be inflated until it is firm enough for the prepreg to be wrapped on directly. The former technique is recommended. The magnitude of the expansion necessary for consolidation is very small; in any event the prepreg plies will not slide over each other, thus restricting a large expansion, and even if they could the possibility of mis-orientation would prohibit this.

After completion of the cure a cooling period is allowed before extraction. The actual time will be dependent upon the mandrel material and its specific heat. For tubes larger than 25 mm diameter, metal mandrels should be hollow for quick cooling and heating. With new mandrels before they are 'worked-in' it has been necessary to cool and then give a short fast re-heat so that the tube expands before the mandrel has a chance to. Extraction is then easier.

After extraction the CFC tubular component is ready for trimming and machining as necessary.

3.7 Pultrusion

The pultrusion process represents an exciting opportunity for CFC component manufacture. Apart from the wide range of profiles possible the process is highly automated without the high labour content that is so dominant in the majority of other manufacturing methods. Pultrusion will play a valuable part in helping CFC to establish itself as an engineering material compared with steel.

The technique basically consists of pulling continuous fibre firstly through a tank containing a resin formulation, followed by a preforming die to remove entrapped air and excess resin from the impregnated fibre and partially to form the required shape. The uncured shape is then drawn into a die which imparts the final shape. Cure is promoted in this die or in a tunnel oven situated at the exit of the die. A diagrammatic representation of the process is depicted in Fig. 3.19(b). There is no typical size, shape or profile of component which the process, or modification thereof, can produce. A wide variety of components which includes I-beams, channel profiles and large area structural panels can be pultruded.

3.7.1 PLANT

It is logical to describe the various pieces of equipment by working systematically along the line, starting at the creel. This can be of simple design and low cost as it is simply a holding rack for the fibre packages. The number of eyelets for guiding the individual tows must be kept to a minimum to reduce abrasion. Tension devices are not required.

94

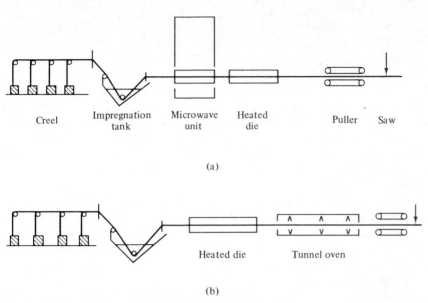

Creel   Impregnation   Microwave   Heated   Puller   Saw
            tank   unit   die

(a)

Heated die   Tunnel oven

(b)

Fig. 3.19 *Schematic layout of two types of pultrusion processes.*

The impregnation tank should be of stainless steel with a long gradual exit slope up which the fibre passes and therefore remains in the flow of resin returning to the bath from the preforming die. In this way, the use of wet-out aids such as fibre spreading bars is eliminated.

The preforming die must be designed carefully and provide gradual changes in section. The surface finish should be smooth, preferably chrome plated to provide a wear-resistant surface. The final shaping die and the method of promoting cure of the resin are the most important aspects of the whole process. A heated die, in which the resin is gelled before final curing in the tunnel oven, is not satisfactory because of the inferior profile and poor surface finish. If the die is maintained at a higher temperature than when curing with a tunnel oven, or if the pulling speed is reduced, cure can be achieved in the last 50 mm of die. This results in an accurate profile with the shape of the die being reproduced exactly with a good surface finish and there is no need for the oven to complete cure. Problems do arise with die curing, but with proper control of the resin formulation and a suitable haul-off unit, it can be done successfully.

Experimental work is in hand to cure the composite with a combination of heat and radio frequency (RF) waves. The use of a heated die alone also creates problems due to a skinning effect, and slow production speeds ensue. The skinning causes internal cracking due to the cure gradient through the section and this has an adverse effect on the mechanical properties. By augmenting the cure with RF waves, the mass is heated from the inside out

95

by molecular excitation. A possible arrangement would be a plastic die tube of PTFE placed after the preforming die, with an RF generator installed around it. Figure 3.19(a) illustrates a typical arrangement. A material such as PTFE is necessary because it is transparent to RF energy and it has a low coefficient of friction which reduces drag. The composite will then pass from the PTFE die into a normal heated steel die which is slightly warmer than itself. This additional heat plus rising self-generated exotherm will complete the cure. The behaviour of CFC under the influence of RF energy has not been fully established and the conductive nature of the fibre could create problems. It is known, however, that polyester resin excitation occurs in the 40 to 70 MHz frequency range and epoxies at 1000 to 2500 MHz and cure can be effected at these frequencies. The combination of RF with a conventional heated die enables large mass sections to be produced at fast speeds. For example, with a glass fibre reinforcement and polyester resin a 25 mm thick section 1 m wide has been run at 2 m per minute on a production basis. With the heated die alone a 25 mm diameter rod can only be run at a speed of 250 mm per minute.

Having discussed the function of dies, their construction must not be neglected. Robust dies are essential because of internal pressure build-up, due to thermal expansion. Adequate bolts are necessary to keep the die closed if it is a split design. The expansion is also the cause of frictional drag, necessitating smooth inside surfaces that are wear resistant. A polished finish which has a 0·03 mm thick layer of chrome plate is recommended. Heating of steel dies should be accomplished with controlled temperature fluid circulating through channels in the die. Silicone oil is an obvious choice and such a system permits temperature change up or down, very quickly.

Moving on from the die assemblies, the next major item of plant in the line is the haul-off unit. It has been stated that this piece of equipment is perhaps the most important of the whole process. Certainly when curing in the die, the ability to control speed and hence cure is dependent upon the pulling unit. It must also be powerful because of the frictional drag mentioned previously. Caterpillar haul-off with a large grip area is the best as it limits the possibility of slippage and ensures full control.

The final piece of equipment in the line is the cut-off saw which is relatively simple in construction and operation. It has to traverse back and forth and when it is cutting it moves along by clamping the CFC pultruded section either side of the intended cut. It is preset to cut off sections, automatically, of the required length.

3.7.2 PULTRUSION PROCEDURE

Having described the plant a resume of the procedures that are adopted when pultruding carbon fibre is called for. With certain grades of fibre, damage due to abrasion can result and a light size is sometimes applied as a protective coating. This is more costly of course, and it should not be necessary if the creel is well designed. Unfilled epoxy resin systems are used predominantly and the incorporation of an internal release agent is a matter of contention at the present time. Small percentages of such chemicals as alcohol phosphates are added to resin formulations when pultruding glass fibre components to reduce the frictional drag between the die and the profile. However, it does not help the composite as a whole to bind together and therefore the smaller the addition the better. This is a problem with CFC in terms of shear strength, even without an internal release agent and on balance the use of internal release agent is not recommended for CFC.

The greatest single cause for rejection of pultruded section is crazing of the surface caused by the frictional forces in the die. The outer surface of the CFC is sheared off during gellation which results in pieces of cured resin sticking to the inside of the die. This scores the surface of the pultrudate as it progresses through the die. The profile deteriorates over a period of time until it finally becomes necessary to stop momentarily to allow the resin particles to cure into the composite. On restarting, they are withdrawn with the section to leave a clean die once more. This process is known as purging and is a necessary evil because it detracts from the continuity of the process as well as incurring the danger of complete jamming of the section in the die if the haul-off unit is not powerful enough to overcome the very high threshold of frictional drag on restarting.

Fibre volume contents as high as 70 per cent are possible with carbon fibre which means that very good mechanical properties can be realized. One present disadvantage is the fact that the fibre orientation can only be unidirectional. This eliminates certain components that are required to function in a multi-directional stress field, but as the available forms of carbon fibre develop, incorporation of off-axis fibre will be possible. It is currently possible, however, to utilize glass fabric or mat to confer transverse strength to CFC pultrusions.

A non-continuous pultrusion process will fulfill an important role in the manufacture of CFC components. It involves the pulling of fibre through an impregnation bath into a closed die and then stopping. The resin is cured usually by the die being heated, after which the die is parted. The component is removed and another batch of impregnated fibre drawn in. This process eliminates the need for an expensive haul-off unit and finely polished dies, but the throughput is greatly reduced. An added advantage is also the capability of forming a component which is contoured, such as an aircraft stiffening stringer.

3.8 Contact Moulding

The majority of glass fibre reinforced plastics components are manufactured by the contact or as it is sometimes called, wet lay-up method. It is well established but labour intensive and prone to variability. However, for the manufacture of large structures there is no alternative method at a comparable cost. The fibre reinforcement has of necessity to be in a form that can be handled, which usually means fabric or chopped strand mat. Carbon fibre has been woven into fabric and chopped strand mats are available, but the early development work in making components by wet lay-up has been disappointing. In the first instance, the fabrics were bulky because of the thick carbon fibre tow. The crimp angle was excessive, resulting in low laminate properties. Attempts to separate the tow into thinner strands thus reducing the crimp and make it more adaptable to the weaving process only resulted in damage to the filaments. However, moderate success in terms of handleability and composite properties has been realized with high modulus or unidirectional fabric constructions, often utilizing a weft of glass fibre.

Chopped strand mats made with various lengths of fibre have pseudo-isotropic properties but because of the difficulties in de-bulking the mat during the wet lay-up process and the high resin content necessary to accomplish this, these have not reached a suitable level. Because of these points the wet lay-up technique incorporating fabric and mat is not an accepted manufacturing method at the present time for CFC components. In the not too distant future this situation will change, helped by the fall in price of the fibre, because of the requirement for large structures such as radar scanners, support beams in civil engineering, boats, and many more. It is probable that the main usage will be as a combination with glass fibre. For example the low stiffness of GRP puts a limit on the size of seagoing boats. CFC incorporated at strategic points would eliminate this restriction.

Aligned short staple mat is a carbon fibre raw material that is showing promise in wet lay-up processing methods. A small percentage of resin is applied so that it binds the individual fibres together in the form of a handleable mat. However, it is still very delicate and requires careful handling. Light, open weave glass scrim cloths can be placed top and bottom of the mat to facilitate the handling. In such a form the mat can be incorporated in a wet lay-up with the fibre orientated in the appropriate direction. Care must be taken when resin is worked into the fibre so that the alignment is not destroyed. The binder is chosen so that it is dissolved by resin which can then wet the fibre.

The addition of carbon fibre to a GRP in such a way can be extremely efficient. Used on the outside of a laminate construction that has to support bending loads, it improves the stiffness markedly, similarly with support columns in compression. These and many more advantages that can be realized promise to stimulate the necessary development work in the next five years.

98

3.9 Non-standard Techniques

When a raw material such as CFC is considered for an application it is unreasonable to expect the buyer of the product to place an order without proof that the material is entirely suitable. This necessitates the manufacture of prototype components for pre-production trials. They must be made at a minimum cost because this may not be recoverable on future sales if the project is not successful. There is also a requirement for components which are of unusual shape or size and as such cannot be manufactured on conventional plant. Situations such as these call for improvizations to the standard techniques described elsewhere in this section.

3.9.1 MOULDS

Epoxy, plaster, and wood moulds are extremely important because they enable the manufacture of a prototype component at a fraction of the cost of a metal mould. They are predominantly used with vacuum bag autoclave moulding operations, although epoxy moulds have been used for compression moulding.

Gypsum is used for plaster moulds and is adaptable for making large moulds which have intricate and irregular shapes. Their ability to be dissolved with water or broken up by incorporating chains which can be pulled out makes them especially useful for moulding partially closed vessels, particularly in filament winding. The contour of the mould is obtained either by the casting of liquid plaster over a master pattern or by building up paste with templates, and wire gauze. A serious disadvantage is the temperature capability which is low. Only resin systems which cure below 130°C can be used.

Epoxy moulds are either cast from the liquid resin or built up with glass fibre reinforcement. Both methods require a master pattern. Figure 3.20 shows a cast epoxy mould and a moulded component. In some cases an original metal component has been adapted to obtain the shape of the mould, with slight modification, where necessary, being machined in afterwards. The choice between a laminating or a casting technique is one of economics, the former being popular for larger moulds or in some cases where the mould requires to be of high strength because high pressures and repeated cycles are anticipated. The resin properties when casting are important. It should be easily poured to avoid air entrapment and the shrinkage on cure must be negligible. The coefficient of thermal expansion must be kept to a minimum and the dimensional stability good at the moulding temperature. A maximum moulding temperature of 160°C with a production life of several thousand mouldings is possible. Epoxies for the manufacture of moulds offer a valuable flexibility in design and manufacturing techniques.

A matured wood must be used when making a mould; machining and carving the shape from a laminated block of, for example 'Stabilite' is quite

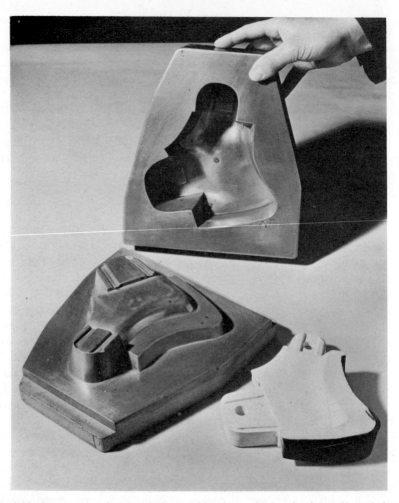

Fig. 3.20 *Cast epoxy resin compression mould and a moulded carbon/glass fibre reinforced component.*

common. Sealing of the mould is necessary before use and this can be done effectively with polyurethane or epoxy lacquers. A maximum cure temperature of 120°C and pressure of 0·6 MN m^{-2} are limitations.

3.9.2 LARGE COMPONENTS

Two particular components that have been required in CFC illustrate extremely well the improvization that is sometimes necessary in manufacture, namely a 3 mm thick panel 3 m × 0·5 m in area, and a 50 mm diameter tube over 6 m long.

With neither an oven nor an autoclave available in which to cure the panel the following technique can be adopted. Lay-up the CFC prepreg in a vacuum bag assembly as shown in Fig. 3.21 and mount on a flat, solid bench. The bag is in fact the heat source by virtue of heating wire incorporated in it. These are commercially available at various ratings to give a required temperature for a given current. A cure of 6 hours at 140°C with vacuum pressure will produce an excellent, well-consolidated panel.

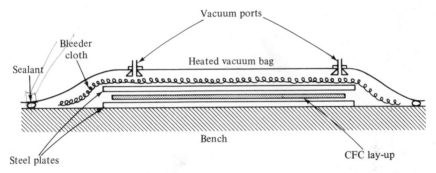

Fig. 3.21 *Lay-up assembly for moulding large panels.*

In the case of the tube two mandrels can be used, each over 3 m long with a loose spigot to join them together. The CFC is hand wrapped and consolidated in 1 m sections on a rolling table. Shrink film is used as a pressure source during cure which can be achieved in a conveyor oven. The two mandrels can be easily extracted from either end of the tube after cure.

It is the intention of this chapter to encourage the design engineer towards unrestricted design thinking. It is hoped that the few examples cited illustrate the wide scope of techniques and shapes possible by manufacturing ingenuity. After all, the standard procedures described elsewhere in this section have become established because of the large demand for the specific types of component that they produce. If a CFC component manufactured by a non-standard technique is successful, then the technique will rapidly be established as standard.

3.10 Adhesive Bonding

Confidence started to grow in the use of adhesives for bonding together load-bearing structures when the aircraft industry adopted the technique twenty-five years ago. Today the techniques and the adhesives represent a technology in their own right.

The processes involved are (a) the surface preparation of the parts to be joined, (b) mixing or cutting to shape the adhesive, (c) application of the adhesive to the parts and, (d) curing the adhesive. The CFC surface should be shot blasted or sanded followed by a degrease ensuring that all traces of mould release agent have been removed. Once prepared, the surface must be protected from contamination by dust, oil, or solvents which can be detrimental to the formation of a good bond. The preparation of metal surfaces such as steel or aluminium is dependent upon the type but acid etching is the preferred method. A typical treatment for aluminium would be:

(a) Vapour degrease, alkaline clean, rinse, and check for water break.
(b) Immerse in sodium dichromate sulphuric acid solution (28·5 g sodium dichromate, 28·5 g concentrated sulphuric acid, make up to 1 litre with distilled water) for 10 minutes at 70°C.
(c) Spray rinse with cold distilled water.

To test the surface for cleanliness after preparation a water drop is deposited on it; if it is contaminated the drop will remain in a globular form. If the surface is clean and suitable for bonding the drop will spread into a thin layer.

3.10.1 ADHESIVES

Thermosetting resins have been found to be the most suitable for bonding CFC to itself and to other substrates. Of the thermosets, epoxies and modified epoxies are preferred and provide a wide range of desirable properties. The following is a list of the commercially available epoxy based adhesives:

epichlorohydrin–bisphenol A	epoxy–nitrile
cycloaliphatic	epoxy–polysulphide
epoxy–polyamide	epoxy–phenolic

A convenient classification is by their temperature capability. The table below indicates this maximum operating temperature for short periods:

standard epoxy	170°C	epoxy–nitrile	120°C
epoxy–polyamide	100°C	epoxy–phenolic	240°C

These adhesives give a high joint strength, can be cured at low temperatures and are suitable for all types of joint. They are available in film or paste form and precatalysed. Certain paste systems, however, can be obtained on a two-component basis so that a catalyst can be chosen to promote a room temperature cure and is thus mixed with the resin prior to application. This can be important when bonding CFC to a material with a different coefficient of linear thermal expansion. Considerable residual stress can result after elevated temperature curing which can lead to severe distortion of the structure on cooling. The film adhesives cannot be recommended too highly because of their strength, handleability and when put onto a fine supporting cloth, their ability to keep adhesive in the joint. A common yet often unrealized reason for premature bonded joint failure is a starved bond line caused by excessive mating pressure before cure. A recommended thickness for CFC joints is 0·1 mm to 0·2 mm to ensure maximum efficiency. Thicker

102

bond lines can be tolerated but not thinner. The price of adhesive films is high compared to the pastes, but this is a cost worth paying for certain jobs where the advantages are realized.

The most common type of joint is a straightforward lap or modifications thereof. Scarf joints are next in importance and are predominantly used in repair work. Straight butt joints are not used unless substantial support straps are added. The basic configurations are depicted in Fig. 3.22. With tubular joints a lap configuration is also favoured; recommended types are shown in Fig. 3.23. Corner and tee joints are critical structural points and the recommended configurations are shown in Fig. 3.24.

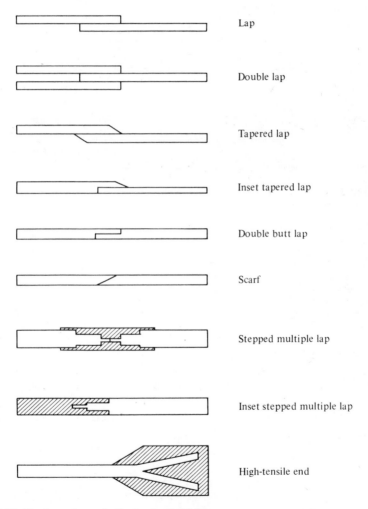

Lap

Double lap

Tapered lap

Inset tapered lap

Double butt lap

Scarf

Stepped multiple lap

Inset stepped multiple lap

High-tensile end

Fig. 3.22 *Configurations of adhesive bonded joints.*

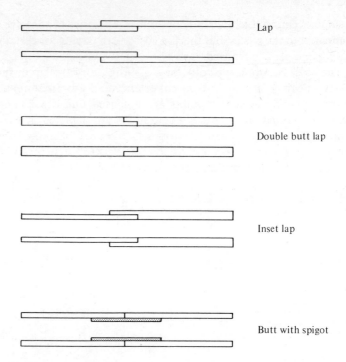

Lap

Double butt lap

Inset lap

Butt with spigot

Fig. 3.23 *Configurations of tubular bonded joints.*

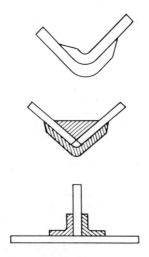

Fig. 3.24 *Recommended corner and tee joints.*

104

With the lap joint the length of overlap should be between 25 mm and 50 mm. Above this the load-carrying ability increases only slightly. The stress concentrations at the ends of the joint are reduced if the lap can be tapered enabling a higher load to be sustained.

3.11 Machining

The following general information applies in all machining operations.

(a) Provide adequate dust extraction to prevent contamination of the machine shop.
(b) Operators should wear face masks to prevent ingestion of the exceedingly fine dust particles.
(c) Coolants. Use general purpose water-soluble oil.
   Advantages
   (i) improved surface finish
   (ii) less heat build-up
   (iii) dust kept to a minimum.
   Disadvantages
   (i) when mixed with machine oils and lubricants, the highly abrasive dust will damage slides, slideways, and other machine parts
   (ii) adverse effect on subsequent bonding operations, which can be partially overcome by degreasing and drying.
(d) Tools. High speed steel tools can be used and give satisfactory results. However, they wear rapidly and tool life between regrinding is short. By using tungsten carbide tools, higher machining speeds are possible and the tools have a longer life.
(e) Special care should be taken to ensure that the CFC component is fully post cured, otherwise the heat generated during machining may cause the resin to soften.

3.11.1 DRILLING

The main difficulty when drilling CFC is to prevent fibre damage and delamination during entry to, and breakthrough from, the composite material. Although it is possible to drill straight through it is a lengthy process due to the very low feed rates possible without damaging the component. The composite material should ideally be sandwiched between waste material, either plastics or metal sheet, and securely clamped. The use of waste material is most important on the breakthrough side of the component. A slow constant feed rate with a woodpecker action is recommended. The use of coolant makes no significant difference in drilling characteristics.

The drills should be two-flute with a high helix angle and run at a speed of 500 surface metres per minute.

3.11.2 TURNING AND BORING

Tungsten carbide tools with a sharp radius give good results and have an acceptable tool life.

Cutting speeds of 120 m per minute at feeds of 0·1 to 0·2 mm per revolution all give good surface finishes with no sign of fibre damage. Using lower cutting speeds than those quoted causes the tool to tear the fibre, giving a very poor surface finish.

3.11.3 MILLING

Radiused end-mills have a longer life than have square end-mills. Using cutters of the 2-flute type results in fibre tearing and delamination, therefore the use of multi-flute cutters is necessary.

Extreme care should be taken when breaking through the edge during face milling, as excessive fibre damage occurs. This can be kept to a minimum if waste material is clamped to the sides of the component.

When machining unidirectional CFC improved surface finish can be obtained if the component is fed to the cutter with the fibres running longitudinally, i.e., mill along the fibres, not across them. A cutting speed of 120 m per minute is optimum.

3.11.4 SURFACE GRINDING

The type of wheel for 'one-off' jobs can be aluminium oxide or silicon carbide (60/40 grit) which give satisfactory results. However, both these standard wheels load up quickly and need frequent dressing to keep them open. Aluminium oxide wheels also dull quickly.

For repeated use the following wheel type is recommended:

> Silicon Carbide (Green Grit)
> Vitrified Bond
> Soft Grade (11 or 1)
> Grit size dependent upon surface finish required.

Components with a large surface area and a thin section may be held to a surface chuck using double-sided adhesive tape. This stage is important; adequate degreasing of both the component and the chuck is required in order to obtain uniform adhesion. Care is needed to ensure that all corners of the component are in contact with the adhesive tape. For end and side grinding, and some types of surface grinding where the use of adhesive is impractical, the components can be clamped in screw jaw vices, so long as adequate support is given to the component to prevent deformation.

A speed of 120 m per minute is recommended.

Very little success has been achieved with punching of CFC because of the tendency to delaminate. Electron discharge and electrochemical machining are under development but the early work has not been encouraging because

106

of the conductive nature of the composite. Of the more advanced machining techniques, laser and ultrasonic methods are showing the most promising results.

Sawing and slitting can be done successfully on silicon carbide or diamond tipped wheels. For accurate cuts a minimum 1 mm wheel thickness is necessary but unfortunately this is a waste of material when profile cutting. Conventional sanding techniques are straightforward: the use of wet and dry paper yields a smooth surface. For even finer mirror finishes honing, lapping, buffing, and polishing techniques can be employed. Coarse and fine grit alumina with water as the carrying medium are used when appropriate.

4. Structural Engineering Design and Applications

R. Tetlow

4.1 Introduction

The design of structures to be fabricated from an anisotropic material is not a new problem. The degree of anisotropy of wood is very similar to that of unidirectional carbon fibre composite and in spite of this and wood's other 'difficult' properties, brittleness, variability, etc., highly successful structures in terms of specific strength and stiffness have been designed. The designer of carbon fibre and, indeed, glass fibre structures is able to utilize the existing knowledge of the behaviour of anisotropic structures to exploit the very high specific strength and stiffness of carbon fibre to the full. Freedom to tailor the material to suit a particular loading combination while permitting greater scope for ingenuity does present an even greater challenge to the designer compared with metal structures.

We shall see that carbon fibre composites offer very high potential weight savings but two major pitfalls must be avoided.

(a) Conventional metal structures cannot simply be made in carbon fibre composites by substitution of the new material. A complete reappraisal of the problem is required and in particular a close look at the design specification in terms of loading, stiffness, size, safety factors, etc.
(b) In spite of the highly desirable properties of carbon fibre composites a true 'composite' structure should be designed, not necessarily one manufactured wholly from carbon fibre composite.

Simplified 'Merit Indices' (e.g., specific modulus = elastic modulus/density) can give a completely false picture of the potential of the material. They can in fact grossly overestimate the worth of the material if the longitudinal properties of a single sheet are used to calculate the merit index. As we shall see later the effectiveness of the material must be determined accurately by taking into account the load level, size of structural member, type of load, etc.

Chapter 1 discusses the available forms of carbon fibre composites, rovings, unidirectional sheet, etc. However, the areas of main interest for the designer

108

are those forms which are essentially unwoven aligned fibres including aligned short fibre mat. The latter behaves similarly to continuous filament although there will obviously be a greater degree of time dependence with resin matrices as the matrix plays a more important role in terms of load transfer.

The price of carbon fibre is discussed elsewhere and it is certainly at this moment very expensive but should reduce in price with increasing utilization. Nevertheless in those structures where reduced structure weight offers direct tradeoff in terms of improved performance, increased payload, etc., considerable savings may be made in terms of reduced direct operating costs. Also in other areas where the reduced structure weight leads to a reduction in the applied loads (e.g., rotating and reciprocating machinery) a twofold gain can be made and indeed in some cases increased productivity can result due to potentially higher machine speeds. Carbon fibre may also have a large market in the sports goods field where its high specific stiffness may be exploited although in some cases the effect on the user may be to introduce an increased performance by psychological rather than structural means.

4.2 Design Philosophy

Unlike design with conventional materials the choice widens to include the correct proportion of filaments to matrix, choice of the matrix material, orientation of individual plies (i.e., lay-up), stacking order, etc. Methods for predicting the mechanical properties of multi-layer composites are available and two basic approaches are possible.

(a) *Netting analysis.* This method, proposed by Cox[1] and adopted by Rothwell[2] for continuous filament composites, assumes that the mechanical properties do not rely on the resin properties but merely on the 'net' of filaments. The resin is merely considered to bind the filaments together but the theory does rely on the fact that each layer of filaments is 'restrained' by the adjacent layers so that strain compatibility is achieved. Although resin strains and stresses are thus implied they are not considered to be contributory to the mechanical properties of the composite. The mechanical properties of the composite are then expressed as a function of the filament properties only. Netting analysis can be used not merely to calculate the elastic and strength properties of a multi-layer composite as suggested by Shibley[3] but also to determine the optimum lay-up for

(i) strength and stiffness under combined loading[4] and
(ii) resistance to buckling[2] under compression or shear loading.

The neglect of the stresses and strains in the resin can, however, lead to serious errors depending on the particular composite as we shall see later.

(b) *Continuum analysis.* Here the actual properties including transverse and shear moduli and strengths of a single layer are used to derive the properties

109

of the total multi-layer composite. Although tedious it does present a true picture of the real performance of the composite.

The effect of varying the fibre volume fraction, V_f, is discussed in chapter 1 but it can be shown that the most effective composite is that containing the maximum volume of fibres but not beyond the limit where the transverse and shear properties are degraded by further increases in the fibre volume fraction. At the moment the working maximum appears to be 60 per cent by volume. Discussion will therefore centre around composites with this amount of fibre. This comment of course only applies to a particular layer of a composite; it does not preclude the design of sandwich panels and beams where the cores may be composed of very lightweight material and the carbon fibre faces should contain the maximum volume of carbon fibre subject to the restraints suggested.

Techniques are available for calculating the properties of a single uni-directional layer[5,6] and while these are effective for the longitudinal properties they are less effective in determining the transverse and shear properties as these latter properties in particular are dependent to a greater degree on the presence of voids and on the closeness of the fibres to each other (contiguity). As a consequence the author feels that a more practical approach is to use the measured properties of the single layer as a basic 'building block'. In the case of 'netting' analysis an average filament strength and stiffness must be used. The approach adopted can be summarized as shown in Fig. 4.1 for both the netting and the continuum method.

4.3 Elastic Properties of Multi-Layer Composites

Accepting that the properties of a single layer can be calculated[5,6] or better still measured for a particular composite the procedure is then as follows.

4.3.1 STRESS-STRAIN RELATIONSHIP

(a) *Continuum analysis.* Each layer has three moduli of elasticity in the direction of the three axes (see Fig. 4.2), one of which is taken to be parallel to the filaments, three moduli of rigidity $G_{\alpha\beta}$, $G_{\beta\gamma}$, $G_{\gamma\alpha}$, and six Poisson's ratios, two associated with each axis, $\mu_{\alpha\beta}$, etc. The latter are not independent but are associated.

$$\left.\begin{array}{l} \dfrac{\mu_{\alpha\beta}}{E_\alpha} = \dfrac{\mu_{\beta\alpha}}{E_\beta} \\[2ex] \dfrac{\mu_{\alpha\gamma}}{E_\alpha} = \dfrac{\mu_{\gamma\alpha}}{E_\gamma} \\[2ex] \dfrac{\mu_{\beta\gamma}}{E_\beta} = \dfrac{\mu_{\gamma\beta}}{E_\gamma} \end{array}\right\} \tag{4.1}$$

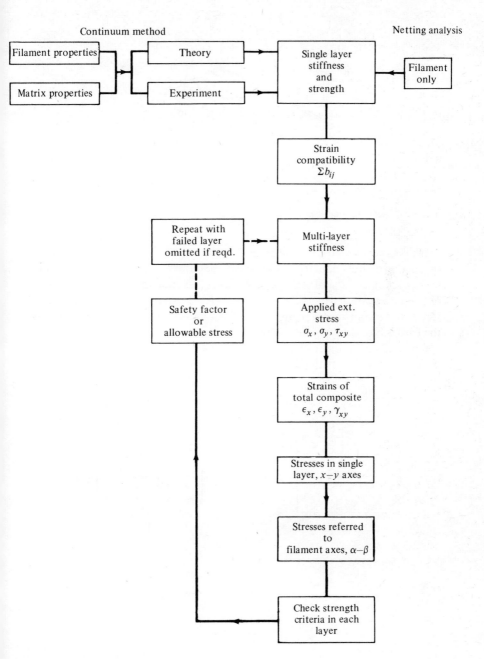

Continuum method

Netting analysis

Fig. 4.1 *Strength and stiffness.*

111

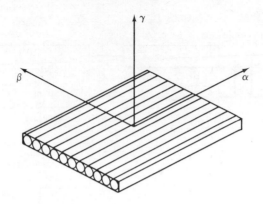

Fig. 4.2 *Single unidirectional ply.*

It thus requires nine independent properties to define the elastic behaviour of an anisotropic material. For sheet materials, it is normal to assume[9] $\sigma_y = \tau_{\alpha\gamma} = \tau_{\gamma\beta} = 0$ and strain ε_γ is ignored, this reduces the required number of elastic constants to E_α, E_β, $G_{\alpha\beta}$, $\mu_{\alpha\beta}$. As we are usually concerned with stress and strain in the plane of the sheet and also with unwoven material, the errors incurred will be negligible.

Consider now the effect of rotating the major axes of the single sheet through an angle θ with respect to the axes of applied stress, x, y (see Fig. 4.3).

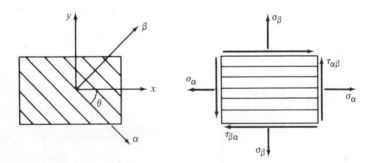

Fig. 4.3 *Single ply.*

Now

$$\begin{bmatrix} \sigma_x \\ \sigma_y \\ \tau_{xy} \end{bmatrix} = [b_{ij}] \begin{bmatrix} \varepsilon_x \\ \varepsilon_y \\ \gamma_{xy} \end{bmatrix} \tag{4.2}$$

112

where b_{ij} = stiffness matrix and

$$b_{11} = [E_\alpha \cos^4 \theta + E_\beta \sin^4 \theta + \sin^2 \theta \cos^2 \theta (2E_\alpha \mu_{\beta\alpha} + 4\lambda G_{\alpha\beta})]/\lambda$$

$$b_{22} = [E_\beta \cos^4 \theta + E_\alpha \sin^4 \theta + \sin^2 \theta \cos^2 \theta (2E_\alpha \mu_{\beta\alpha} + 4\lambda G_{\alpha\beta})]/\lambda$$

$$b_{33} = [\sin^2 \theta \cos^2 \theta (E_\alpha + E_\beta - 2E_\alpha \mu_{\beta\alpha}) + \lambda G_{\alpha\beta}(\cos^2 \theta - \sin^2 \theta)^2]/\lambda$$

$$b_{21} = b_{12} = [\sin^2 \theta \cos^2 \theta (E_\alpha + E_\beta - 4\lambda G_{\alpha\beta}) + E_\alpha \mu_{\beta\alpha}(\cos^4 \theta + \sin^4 \theta)]/\lambda$$

$$b_{31} = b_{13} = [\sin^3 \theta \cos \theta (E_\beta - E_\alpha \mu_{\beta\alpha} - 2\lambda G_{\alpha\beta}) - \sin \theta \cos^3 \theta$$
$$(E_\alpha - E_\alpha \mu_{\beta\alpha} - 2\lambda G_{\alpha\beta})]/\lambda$$

$$b_{32} = b_{23} = [\sin \theta \cos^3 \theta (E_\beta - E_\alpha \mu_{\beta\alpha} - 2\lambda G_{\alpha\beta}) - \sin^3 \theta \cos \theta$$
$$(E_\alpha - E_\alpha \mu_{\beta\alpha} - 2\lambda G_{\alpha\beta})]/\lambda$$

$$\lambda = 1 - \mu_{\beta\alpha} \mu_{\alpha\beta}$$

and also by inversion

$$\begin{bmatrix} \varepsilon_x \\ \varepsilon_y \\ \gamma_{xy} \end{bmatrix} = [a_{ij}] \begin{bmatrix} \sigma_x \\ \sigma_y \\ \tau_{xy} \end{bmatrix} = [b_{ij}]^{-1} \begin{bmatrix} \sigma_x \\ \sigma_y \\ \tau_{xy} \end{bmatrix} \qquad (4.3)$$

$[a_i]$ = compliance matrix

and hence

$$\left. \begin{aligned} E_x &= \frac{1}{a_{11}} = \frac{b_{11}b_{22} - b_{12}^2}{b_{22}} \\[2mm] E_y &= \frac{1}{a_{22}} = \frac{b_{11}b_{22} - b_{12}^2}{b_{11}} \\[2mm] G_{xy} &= \frac{1}{a_{33}} = b_{33} \\[2mm] \mu_{xy} &= -\frac{a_{12}}{a_{11}} = \frac{b_{12}}{b_{22}} \\[2mm] \mu_{xy} &= -\frac{a_{12}}{a_{22}} = \frac{b_{12}}{b_{11}} \end{aligned} \right\} \qquad (4.4)$$

Figure 4.4 shows the effect on a single layer of a typical carbon fibre composite.

The behaviour of laminated plates can now be determined by equating the strains in each layer such that compatibility is achieved. Consider first a total composite made from an infinite number of layers evenly dispersed through the composite such that the effect of uneven stacking may be ignored.

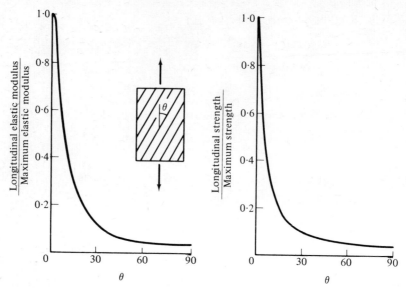

Fig. 4.4 *Single layer—orientation effects.*

The properties of the total composite can be determined by summing the stiffness matrices of each layer proportional to the number of plies in a particular direction thus

$$[b_{ij}]_{\text{comp}} = \frac{1}{t} \sum_{k=1}^{k=n} b_{ij_k} \cdot t_k \qquad (4.5)$$

where k = single ply layer.

If the number of layers is small then the effect of the stacking sequence must be considered as unsymmetric lay-ups can produce plate twisting and/or bending when in-plane loads are applied and a different procedure must be adopted as follows.

$$\text{Load/in} = N = \int_{-t/2}^{t/2} \sigma_k \cdot \mathrm{d}z \qquad (4.6)$$

$$\mathrm{d}z = \text{thickness of } k\text{th layer}$$

$$\text{bending moment/in} = M = \int_{-t/2}^{t/2} \sigma_k \cdot z \cdot \mathrm{d}z \qquad (4.7)$$

and

$$\varepsilon = \text{in-plane strain}$$

$$x = \text{bending curvature}$$

114

therefore

$$\text{total strain in } k \text{ layer} = \varepsilon + z \cdot x \tag{4.8}$$

and

$$\begin{bmatrix} N \\ M \end{bmatrix} = \begin{bmatrix} B & \vline & C \\ \hline C & \vline & D \end{bmatrix} \begin{bmatrix} \varepsilon \\ x \end{bmatrix} \tag{4.9}$$

and the elements of the partitioned matrix can be defined as

$$[B], [C], [D] = \int_{-t/2}^{t/2} (1, z, z^2)[b_{ij}]_k \, dz \tag{4.10}$$

where $[b_{ij}]_k$ has been previously defined for the kth layer; the matrix equation can then be inverted to determine the elastic properties.

Strains induced due to thermal effects can also be dealt with in a similar manner.[5]

The elastic properties for a three-ply system (Fig. 4.5) can be seen in Figs. 4.7, 4.8, and 4.9 for a complete range of cross ply angles, θ, and cross ply ratios, R, where $R = $ number of cross plies/total plies.

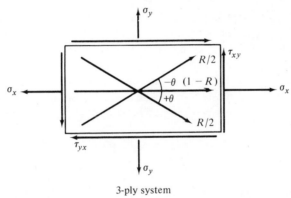

3-ply system

Fig. 4.5 *Sign convention.*

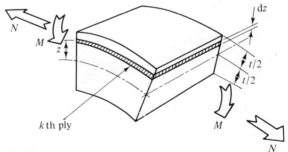

Fig. 4.6 *Ply stacking.*

115

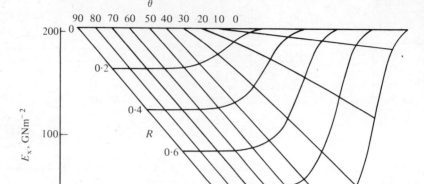

Fig. 4.7 *Longitudinal elastic modulus, E_x.*

For practical purposes it is sufficient at the initial design stage to assume that the lay-up will be symmetrical and the layers evenly dispersed. At a later stage the effect of the stacking order can be introduced as a refinement.

The use of a very simple computer program enables a whole range of multilayer composites to be analysed and the results may then be scanned to determine the optimum lay-up for the desired ratio of elastic properties (see section 4.13). Poisson's ratio is not plotted but it can be shown that with

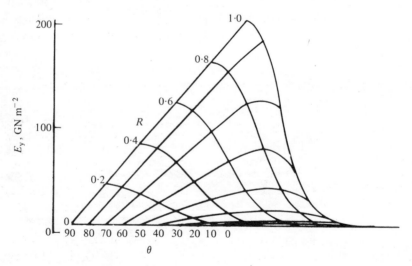

Fig. 4.8 *Transverse elastic modulus, E_y.*

116

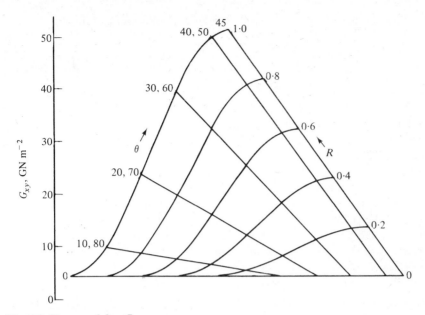

Fig. 4.9 *Shear modulus, G_{xy}.*

$R = 1\cdot0$ (i.e., all angle plies) and for small cross ply angles, θ, that $\mu_{xy} > 1\cdot0$ (this is verified by Cox[1]) and this would be accompanied by an appropriate Poisson's ratio through the thickness of the laminate in order that the volume of the plate remained constant. Of particular interest is the high shear modulus which can be achieved by cross plying at $\pm45°$. Table 4.1 shows the possible weight savings for a slow-running, heavily-loaded, thick-wall torque tube. The tube could be manufactured using existing technology and equipment as a filament wound tube.

Table 4.1 *Comparative weights of torque tubes (slow running—thick wall)*

Material	Allowable Shear Stress $kN\,m^{-2}$	Shear Modulus $\times 10^6\,kN\,m^{-2}$	Comparative Weight Based on Strength	Stiffness
HM CF ($\pm45°$)	372 300	52·54	1	1
Aluminium Alloy, L73	234 400	26·54	2·65	3·30
High Strength Steel Tube, T2	772 200	77·22	2·40	3·40

Failure criteria for anisotropic materials are discussed in detail by Chamis[7] and Tsai.[8] The most effective theory for carbon fibre appears to be that proposed by Hill[10] although this is strictly a yield criterion.

117

The Hill criterion can be expressed for a single layer as follows.

$$K = \frac{X}{\sqrt{[\sigma_\alpha^2 - \sigma_\alpha \cdot \sigma_\beta + (X^2/Y^2) \cdot \sigma_\beta^2 + (X^2/S^2) \cdot \tau_{\alpha\beta}^2]}} \quad (4.11)$$

where X, Y, and S are the measured or predicted longitudinal, transverse, and shear strengths, respectively, and the applied stresses σ_α, σ_β, and $\tau_{\alpha\beta}$ are referred to the major axes of the layer. Tsai[10] has modified this criterion by multiplying the second term in the denominator by (X/Y) as a result of his tests on glass fibre reinforced plastics. There appears, however, to be no justification for this for carbon fibre reinforced plastics and indeed it appears somewhat dubious for 0–90° glass fibre reinforced plastic lay-ups.

If the elastic properties of the multi-layer plate are known then the strains in each layer for a given ratio of applied stresses, σ_x, σ_y, and τ_{xy} can be calculated and hence the stresses in each individual layer and these can then be referred to the major axes of the layer.

Hence

$$\begin{bmatrix} \varepsilon_x \\ \varepsilon_y \\ \gamma_{xy} \end{bmatrix}_{comp} = [a_{ij}]_{comp} \begin{bmatrix} \sigma_x \\ \sigma_y \\ \tau_{xy} \end{bmatrix} \quad (4.12)$$

and

$$\begin{bmatrix} \sigma_x \\ \sigma_y \\ \tau_{xy} \end{bmatrix}_{layer} = [b_{ij}]_{layer} \begin{bmatrix} \varepsilon_x \\ \sigma_y \\ \gamma_{xy} \end{bmatrix}_{comp} \quad (4.13)$$

$$= [b_{ij}]_{layer} [a_{ij}]_{comp} \begin{bmatrix} \sigma_x \\ \sigma_y \\ \tau_{xy} \end{bmatrix}_{comp} \quad (4.14)$$

and then as the stresses in the plate major axes, α, β are required

$$\begin{bmatrix} \sigma_\alpha \\ \sigma_\beta \\ \tau_{\alpha\beta} \end{bmatrix}_{layer} = [mn] [b_{ij}]_{layer} [a_{ij}]_{comp} \begin{bmatrix} \sigma_x \\ \sigma_y \\ \tau_{xy} \end{bmatrix}_{comp} \quad (4.15)$$

where $mn = \begin{bmatrix} \cos^2 \theta & \sin^2 \theta & -2 \sin \theta \cos \theta \\ \sin^2 \theta & \cos^2 \theta & 2 \sin \theta \cos \theta \\ \sin \theta \cos \theta & -\sin \theta \cos \theta & (\cos^2 \theta - \sin^2 \theta) \end{bmatrix}$

These stresses can then be substituted into the chosen strength criterion and the strength factor, K, determined. Failure will occur in one of the layers and in some cases (e.g., simple tension) it would then be possible to carry out a further analysis of the remaining layers excluding the failed layer in order to determine the final strength. In most cases this is only of academic interest since the primary failure would be considered to be the design stress level. For instance in those areas where transverse stress, σ_y, or shear stress, τ_{xy}, is present in addition to longitudinal stress then the primary failure would immediately precipitate total failure. A sample calculation is included in section 14.13 and also the results are shown for a range of laminates composed of high modulus carbon fibre in simple tension in Fig. 4.10.

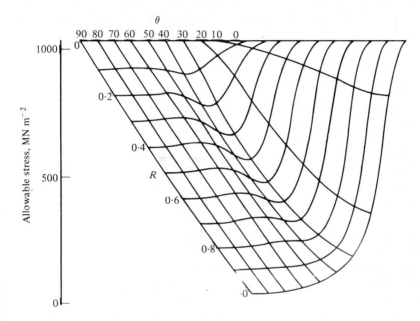

Fig. 4.10 *Tensile strength HMCF.*

Also the stress–strain curves in simple tension for a 0–90° cross ply composite based on high modulus and high strength carbon fibre reinforced plastic are shown in Fig. 4.11. It will be noted that the primary failure in the high strength variety is predicted in the cross ply unlike the high modulus where the primary failure is in the axial ply.

Again by use of a computer program for the desired ratio of applied stresses, σ_x, σ_y, and τ_{xy} the results may be scanned to determine the best lay-up.

(b) *Netting analysis.* Each layer is considered to have only axial strength and stiffness proportional to the fibre volume fraction.

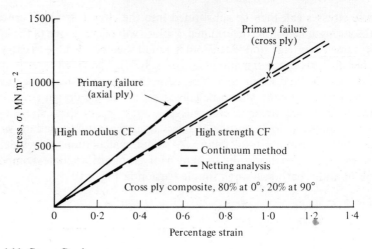

Fig. 4.11 *Stress-Strain.*

Thus

$$\left.\begin{array}{l} X = V_f \cdot \sigma_f \\ E_x = V_f E_f \end{array}\right\} \tag{4.16}$$

where

$$\sigma_f = \text{filament strength}$$

$$E_f = \text{filament modulus}$$

$$Y = 0 = S$$

and Poisson's ratio is assumed to be zero, therefore $\lambda = 1$.

As the layer is only capable of resisting loads in the direction of the filament then for any combined two-dimensional stress system at least a three-fibre system is required.[4]

Rothwell[2] has investigated a symmetric four-fibre system (Fig. 4.12) and by substitution of the above into eq. (4.2).

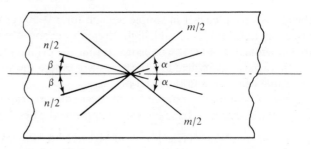

Fig. 4.12 *Symmetric four-ply lay-up.*

120

Then

$$b_{11} = V_f \cdot E_f(m \cos^4 \theta_m + n \cos^4 \theta_n)$$

$$b_{22} = V_f \cdot E_f(m \sin^4 \theta_m + n \sin^4 \theta_n) \qquad\qquad (4.17)$$

$$b_{12} = b_{21} = b_{33} = V_f \cdot E_f(m \cos^2 \theta_m \cdot \sin^2 \theta_m + n \cos^2 \theta_n \sin^2 \theta_n)$$

and

$$b_{13} = b_{31} = b_{23} = b_{32} = 0$$

m, n = proportion of plies in m and n layers respectively. $m + n = 1$.

The analysis is then similar to that used for the continuum analysis except that there are only stresses in the layer in the fibre direction. Whilst this simplifies the procedure considerably and leads to a much simpler solution for desired stiffnesses and strengths it can lead in certain cases to serious errors. Consider a simple 0–90° cross ply arrangement composed of high strength carbon fibre.

Then

$$\text{axial strength} = V_f \cdot \sigma_f \cdot (1 - R) \qquad\qquad (4.18)$$

The result can be compared with the continuum analysis in Fig. 4.11 and it can be seen that the continuum analysis predicts failure in the cross ply at a stress level below that predicted by netting analysis. This premature failure of the cross ply could of course be prevented by a more flexible, stronger matrix.

The application of netting analysis to filaments with an even lower modulus but of similar strength (e.g., glass fibre filaments) accentuates the problem.

This problem is not present with the high modulus carbon fibre reinforced plastics and netting analysis is adequate.

Harris[4] has determined the optimum fibre orientations for three-fibre systems for asymmetric loadings and these will certainly be satisfactory for high modulus carbon fibre reinforced plastics but should be used with caution for other more flexible composites.

4.4 Design of Compression Panels

4.4.1 SIMPLY SUPPORTED PLATES

The theoretical analysis of the buckling of anisotropic plates in compression is well established[12,13] and in the case of glass fibre reinforced plastics and plywood panels has been verified experimentally.[14,15] Tests carried out at Cranfield Institute of Technology[16,17] indicate that the theory is applicable to carbon fibre composites.

An expression for the buckling stress in compression of an orthotropic panel (where major axes are parallel and perpendicular to the direction of loading), has been derived[12] as shown below.

$$\text{Buckling stress, } \sigma_{\text{CR}} = \frac{2\pi^2}{b^2 t} \left[\sqrt{(D_1 D_2)} + D_3 \right] \qquad (4.19)$$

where

b = plate width

t = plate thickness

D_1 = flexural rigidity corresponding to bending moment M_x (see Fig. 4.13)

$\quad = (EI)_x / \lambda_{xy}$

D_2 = flexural rigidity corresponding to bending moment M_y

$\quad = (EI)_y / \lambda_{xy}$

$D_3 = \frac{1}{2}(\mu_{xy} \cdot D_2 + \mu_{yx} \cdot D_1) + 2(GI)_{xy}$

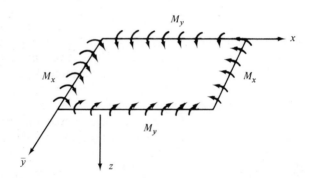

Fig. 4.13 *Plate bending.*

$(GI)_{xy}$ is the average torsional rigidity of the plate.

$\lambda_{xy} = 1 - \mu_{xy} \cdot \mu_{yx}$

For a thin plate, thickness t, these stiffnesses may be expressed as

$$D_1 = E_x \cdot t^3 / 12 \lambda_{xy}$$
$$D_2 = E_y \cdot t^3 / 12 \lambda_{xy} \qquad (4.20)$$
$$D_3 = t^3(\mu_{xy} \cdot E_y + \mu_{yx} \cdot E_x)/24\lambda_{xy} + t^3 G_{xy}/6$$

Then substituting (4.20) into (4.19) gives

$$\sigma_{CR} = \frac{\pi^2}{6\lambda_{xy}}[\sqrt{(E_x . E_y)} + \mu_{xy} . E_y/2 + \mu_{yx} . E_x/2 + 2\lambda_{xy}G_{xy}]\left[\frac{t}{b}\right]^2 \quad (4.21)$$

or

$$\sigma_{CR} = E_{x0}(K')\left(\frac{t}{b}\right)^2 \quad (4.22)$$

$$E_{x0} = \text{longitudinal modulus of a single layer}$$

and

$$K' = \frac{1}{E_{x0}} . \frac{\pi^2}{6\lambda_{xy}}[\sqrt{(E_x . E_y)} + \mu_{xy}E_y/2 + \mu_{yx}E_x/2 + 2\lambda_{xy}G_{xy}] \quad (4.23)$$

The expression K' has been evaluated for plates with a three-fibre system and balanced lay-up and the results are shown in Fig. 4.14.

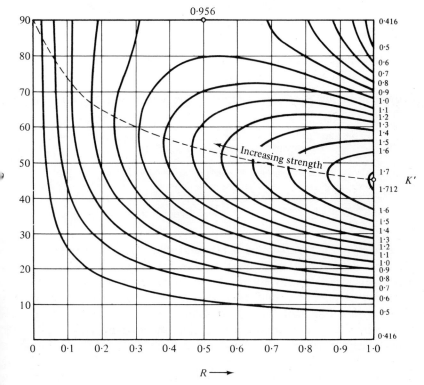

Fig. 4.14 *Simply supported compression panel, buckling coefficients.*

123

It can be seen that the maximum value occurs when $R = 1$, $\theta = 45°$ (i.e all cross plied at $\pm 45°$) but for this lay-up the maximum allowable compressive stress is only $103\cdot5$ MN m^{-2} (Fig. 4.15) compared with a maximum stress of 828 MN m^{-2} for a single unidirectional layer.

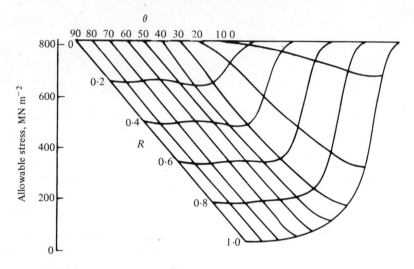

Fig. 4.15 *Compressive strength—HMCF.*

By rearranging eq. (4.22) in terms of the applied load, P, and the panel width, b, then

$$\text{optimum stress, } \sigma_{\text{opt}} = (E_{x0} \cdot K')^{1/3} \left(\frac{P}{b^2}\right)^{2/3} \qquad (4.24$$

and

$$\left(\frac{P}{b^2}\right) = \text{structural index}$$

From eq. (4.24) it is apparent that the maximum stress is not only a function of the material properties but also the structural index. Hence as the structural index is increased it will be necessary to introduce an increasing number of axial plies or plies at a small angle in order to increase the optimum stress level. By superimposing lines of constant stress on the graph for buckling coefficient K' a 'path to follow' from the optimum lay-up ($R = 1\cdot0$, $\theta = 45°$ may be determined and is shown as a broken line in Fig. 4.14. The optimum stress versus the loading index can then be determined and is shown in Fig 4.16. When the structural index is known the optimum stress may be read from the graph and the lay-up obtained by reference to Fig. 4.14.

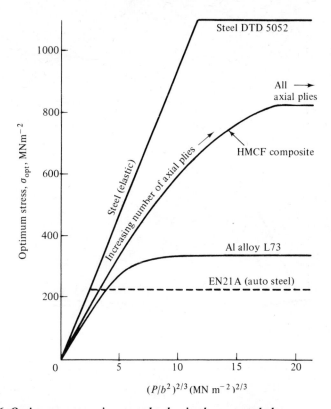

Fig. 4.16 *Optimum compression stress level—simply supported plate.*

Figure 4.16 also shows the attainable stress levels for steel and aluminium for comparison. The weight of aluminium, steel, and woven glass fibre reinforced plastic compression panels are compared with carbon fibre in Fig. 4.17.

The buckling chart (Fig. 4.14) having been derived, this can now be used to determine the buckling stress of plates with other edge conditions. Wittrick[18] has suggested that the curves[19] which give a buckling coefficient, K, versus panel aspect ratio, a/b, for various plate edge conditions may be used by substituting a value

$$N = \frac{a}{b} \cdot \left(\frac{D_2}{D_1}\right)^{1/4} \quad \text{for } a/b \tag{4.25}$$

The buckling stress,

$$\sigma_{CR} = E_{x0} \frac{K_{R.Ae.S}}{3 \cdot 62} \cdot K'(t/b)^2 \tag{4.26}$$

125

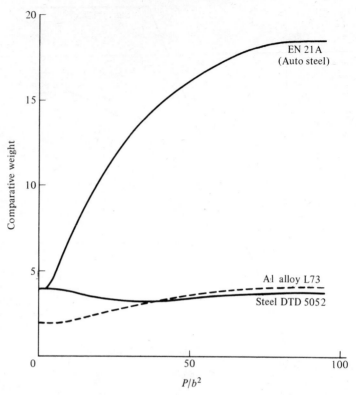

Fig. 4.17 *Comparative weight of simply supported plate in compression.*

and $K_{R.Ae.S}$ = buckling coefficient for isotropic plate[19] for appropriate edge conditions.

Netting analysis leads to similar results for the buckling coefficient chart (Fig. 4.14) except that the values on the axes $R = 0$ and $\theta = 0$ and also when $R = 1$, $\theta = 90°$ are zero. In addition the 'optimum' lay-up, i.e., $R = 1.0$, $\theta = 45°$ has no axial strength and therefore in any real situation axial plies would obviously be required.

4.4.2 COLUMNS

Of particular interest in structural design are wide columns composed of the plate elements previously discussed. Optimum design procedures for wide compression panels are almost invariably based on the criteria that all modes of buckling occur simultaneously at the design load.[20] The procedure adopted for carbon fibre composites is similar in that local and long wave instability modes are assumed to be coincident. The types of panel considered do not buckle in the torsional mode and in fact panels prone to this mode of buckling appear to be less attractive when constructed from unidirectional fibre reinforced plastics.

The design of a corrugated compression panel is first considered and the design of other panel shapes can be derived from the results.

The modes of buckling of a corrugated compression panel to be anticipated are as follows:

a) local buckling of plate elements
b) long wave buckling of whole panel.

For (a) the plate is assumed to be simply supported and infinitely long, although if necessary the interaction effect with the adjacent plates could be included and for (b) ends of the panel are assumed to be simply supported and the effect of edge supports negligible, i.e., wide column.

In this initial analysis the column is assumed to be made from a constant thickness sheet (Fig. 4.18) where

$$t_c = t_f = t, \qquad b_c = b_f = b \tag{4.27}$$

Then local buckling stress of plate,

$$\sigma_{LI} = E_{x0}(K') \left(\frac{t}{b}\right)^2 \tag{4.28}$$

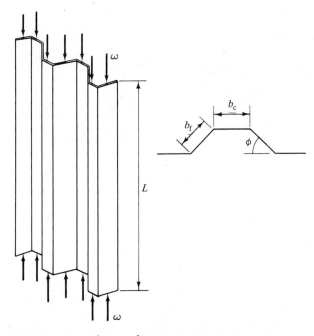

Fig. 4.18 *Corrugated compression panel.*

127

where E_{x0} and (K') are defined previously. Long-wave buckling stress o
panel,

$$\sigma_{LW} = \frac{\pi^2 E_x \cdot I}{L^2 \bar{t}} \tag{4.29}$$

where

$$\bar{t} = \frac{2t}{(1 + \cos \phi)} \tag{4.30}$$

$$I = \left[2bt \left(\frac{h}{2} \right)^2 + \frac{2t}{\sin \phi} \cdot \frac{h^3}{12} \right] \cdot \frac{1}{2b(1 + \cos \phi)} \tag{4.31}$$

for $0 < \phi < 120°$

$$h = b \sin \phi \tag{4.32}$$

Substituting eq. (4.32) into eq. (4.31) we obtain

$$I = \frac{1}{3} \frac{b^2 t \sin^2 \phi}{(1 + \cos \phi)} \tag{4.33}$$

and by substituting eqs. (4.33) and (4.30) into eq. (4.31) we obtain

$$\sigma_{LW} = \frac{\pi^2 \cdot E_x \cdot \sin^2 \phi}{6} \cdot \frac{b^2}{L^2} \tag{4.34}$$

Now applied stress,

$$\sigma_A = \frac{\omega}{\bar{t}} = \frac{\omega(1 + \cos \phi)}{2t} \tag{4.35}$$

$$\omega = \text{load/unit width}$$

As all buckling modes occur simultaneously

$$\sigma_{LI} = \sigma_{LW} = \sigma_A = \sigma \tag{4.36}$$

and by examining eqs. (4.28), (4.29), and (4.34) we can show

$$\sigma^4 = \sigma_{LI} \cdot \sigma_{LW} \cdot \sigma_A^2 = \frac{\pi^2 \cdot E_x \sin^2 \phi}{6} \cdot \frac{b^2}{L^2} \cdot E_{x0}(K') \left(\frac{t}{b} \right)^2 \cdot \frac{\omega^2}{4t^2} (1 + \cos \phi)^2$$

Rearranging, this becomes

$$\sigma^4 = \frac{\pi^2}{24} [E_x \cdot E_{x0}(K')] [\sin^2 \phi (1 + \cos \phi)^2] \left[\frac{\omega}{L} \right]^2 \tag{4.37}$$

or

$$\sigma^4 = \frac{\pi^2}{24} \cdot E_{x0}^2 \cdot Z \cdot M \cdot \left[\frac{\omega}{L} \right]^2 \tag{4.38}$$

128

where

$$Z = \frac{E_x \cdot E_{x0} \cdot K'}{E_{x0}^2}$$

$$M = \sin^2 \phi (1 + \cos \phi)^2$$

By examining eq. (4.38) it can be seen that the parameters for material properties, Z, and cross sectional geometry, M, are independent and for a given loading index, ω/L may be maximized separately in order to achieve the maximum stress, σ, and hence the lightest panel. The maximum value of M occurs when $\phi = 60°$ and substituting this value into eq. (4.38) yields

$$\sigma = 0.914(E_{x0}^2 \cdot Z)^{1/4}(\omega/L)^{1/2} \tag{4.39}$$

The material property parameter, Z, has been evaluated and the results are shown in Fig. 4.19. It can be seen that the maximum value of Z occurs when $R = 0.33$, $\theta = 55$. This 'optimum' orientation will only apply up to the stress level that can be achieved by this particular configuration. In order to increase the stress level (i.e., at a higher loading index, ω/L) then the lay-up

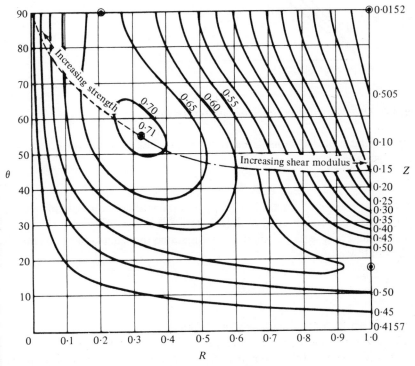

Fig. 4.19 *Corrugated compression panel, buckling chart.*

must be altered and the 'path to follow' from the optimum is obtained as before by superposition of lines of constant allowable stress and the path is shown in Fig. 4.19. By following this path a graph of optimum stress level, σ_{opt} versus loading index, ω/L may be constructed and is shown in Fig. 4.20 in comparison with aluminium alloy.

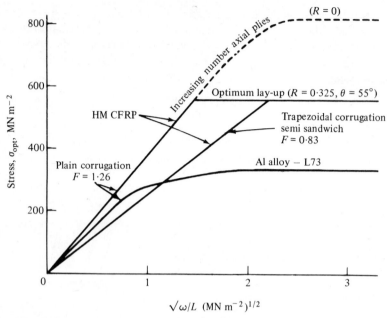

Fig. 4.20 *Optimum stress—corrugated compression panel.*

When the material considered is isotropic, i.e., $E_x = E_{x0} = E$, etc., eq. (4.39) becomes

$$\sigma = F\sqrt{E}\sqrt{(\omega/L)} \qquad (4.40)$$

where F = Efficiency Factor ($= 1.26$ for corrugated panel) and this agrees with the expression obtained by Emero and Spunt.[21]

For other cross sectional shapes it is suggested that eq. (4.40) can be modified as follows

$$\sigma = 0.914(E_{x0}^2 \cdot Z)^{1/4} \left(\frac{\omega}{L}\right)^{1/2} \cdot \frac{F'}{1.26} \qquad (4.41)$$

where F' = Efficiency Factor of panel (table 4.2).

The method suggested applies to panel shapes shown but will provide a useful guideline for other geometries prone to failure by torsional instability providing this is checked. An example of the method is shown in ref. 23.

130

Table 4.2 *Efficiency factors for wide columns*[21]

Type		F
Plain corrugation		1·26
Trapezoidal corrugated, semi sandwich		0·83
Truss core, semi sandwich		0·83
Semi trap. corrugated semi sandwich		0·85
Top hat stiffened		0·96
Truss core corrugation		1·07
Semicircle corrugation		0·84
Truss core sandwich		0·78

$$\sigma_{\text{opt}} = F\sqrt{E}\ \sqrt{\frac{\omega}{L}}$$

Richards[19] has used netting analysis and a simplified lay-up to achieve similar results but has also included the effect of varying the filament volume ratio, V_f. Within the limitations of netting analysis previously discussed the answers are comparable.

4.5 Design of Shear Panels

The buckling, under shear loading, of orthotropic panels is detailed in reference 12. Two types of panel have been considered here; (a) a plain panel and (b) a corrugated panel composed of elements of type (a). The results of the analysis for type (a) may be applied to those panels supported by stiffeners which do not directly resist the shear loading and type (b) may be used for panels similar to corrugations (i.e., with shear-carrying stiffeners).

(a) *Plain panels.* Consider the buckling of a plain panel, infinitely long, all sides simply supported:

$$\text{Buckling shear flow, } q_{\text{CR}} = \frac{4K(D_2 D_3)^{1/2}}{b^2} \quad \text{for } \theta < 1\cdot 0$$

$$= \frac{4K(D_1 D_2^3)^{1/4}}{b^2} \quad \text{for } \theta > 1\cdot 0$$

where $\theta = (D_1 D_2)^{1/2}/D_3$.

131

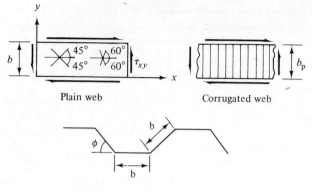

Plain web   Corrugated web

Corrugation geometry

Fig. 4.21 *Shear panels.*

By substituting the appropriate flexural rigidities in terms of the plate thickness as in section 4.4 it can be shown that the maximum value of buckling shear stress, τ_{CR}, occurs when $\theta = 60°$ (i.e., all angle plies, see Fig. 4.21). However this lay-up does not have the maximum shear strength which occurs when all plies are at $\pm 45°$. The choice of lay-up within this range will depend upon the shear loading index, $(q/b)^{2/3}$ and may be determined in a manner similar to that used in section 4.4. Figure 4.22 shows the two extremes for a HM composite: for low load levels a $\pm 60°$ lay-up is required and for high load levels a $\pm 45°$.

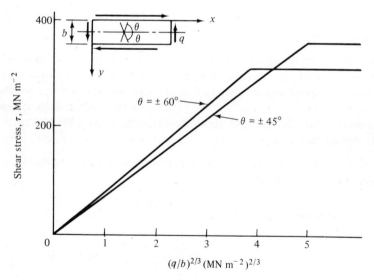

Fig. 4.22 *Optimum shear stress—HMCFRP.*

132

(b) Corrugated shear panels. A corrugated shear panel is another form of orthotropic plate except that in this case the plate elements may also be orthotropic. Two forms of instability are possible:

(a) local buckling of the crest or flank (Fig. 4.21) and
(b) overall plate buckling.

In order to design a minimum weight panel for a given shear loading, q, and depth, b_p, the assumptions are:

(a) local and overall buckling occur simultaneously
(b) flank and crest have the same thickness (this ratio could of course be varied to produce a family of designs).
(c) panel is infinitely long, all sides simply supported.

The analysis has been carried out by the author[23] and like the compression panel it can be seen that:

$$\tau_{opt} = f(\text{material properties, panel geometry}) . (q/b_p)^{1/2}$$

The optimum corrugation angle, $\phi = 60°$ and the variation of stress with loading index is shown in Fig. 4.23 for a $\pm45°$ lay-up. This particular graph

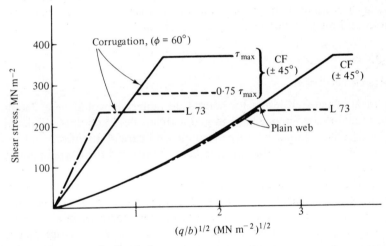

Fig. 4.23 *Shear stress–loading index.*

is only intended as a guide and the whole range of lay-ups should be fully investigated. The graph does show the efficiency of CF compared with aluminium alloy although in this case its relative efficiency is not as high as that of a plain CF panel (Fig. 4.22).

The weight of CF panels compared with aluminium alloy are shown in Fig. 4.24.

133

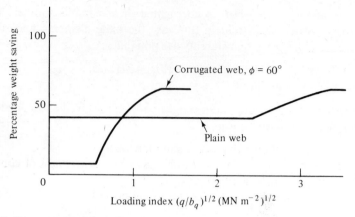

Fig. 4.24 *Percentage weight saving (CF v aluminium alloy)—shear webs.*

4.6 Buckling under Combined Loading

Methods of analysis are available[14] but a family of design charts would be needed to cover all combinations of shear and compression, etc. It is suggested that the analysis be performed in a similar way to the foregoing method for a specific case rather than for the whole range. Even then the calculations are tedious and must be done for a specific material.

4.7 Joints

4.7.1 INTRODUCTION

The ideal structural design for composite materials would be a one-piece structure fabricated in one large mould with, in effect, no joints. While this is a pipe dream it should be borne in mind and careful attention paid to the design concept in order that the number of joints and the loads imposed on them can be minimized.

Joints in any material almost inevitably lead to a weight penalty but unlike composites most metallic materials have more forgiving properties in that they will yield under load and have comparable properties in all directions. The weight penalties of joints in composite structure will be greater than those for metals if the number required is the same. The manufacture of a composite structure does, however, lend itself to simple local reinforcing and also a reduction in the number of joints and as a result the weight penalty can be considerably reduced. An excellent example of the latter can be seen in the construction of the wings of the increasing number of GFRP sailplanes; the spar caps which resist the very high bending moments are composed of uni-directional glass rovings without any joints in the axial sense. The load transfer is achieved not by cutting the rovings by bolting or attempting to

134

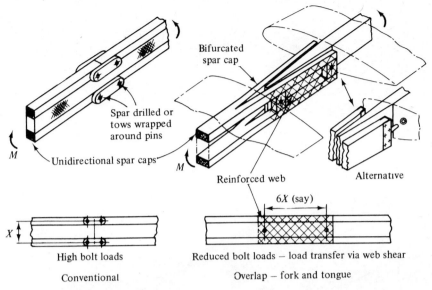

Bifurcated
spar cap

Spar drilled or
tows wrapped
around pins

M

Unidirectional spar caps

M

Reinforced web

Alternative

$6X$ (say)

X

High bolt loads

Reduced bolt loads — load transfer via web shear

Conventional

Overlap — fork and tongue

Fig. 4.25 *Wing spar joint—Sailplane.*

wrap them around pins but by redistribution in shear over a long distance (Fig. 4.25).

4.7.2 JOINT CLASSIFICATION

There are several types of possible joints for composites as follows:

(a) *Mechanical.* Simple bolted, riveted, and pinned, single and double lap joints. Various types of 'wedge' joints relying on the relatively high flatwise compressive strength of the laminate. Also in this category are joints where load transfer is achieved via tows of filament wrapped around pins (Fig. 4.26).

(b) *Bonded.* Simple bonded, single or double lap joints and scarf or stepped scarf joints (Fig. 4.27). Included in this category are joints with interleaved shims of reinforcing material (e.g., glass or asbestos mat cloth, titanium, steel, CF mat, etc.).

(c) *Mixed.* Joints using a combination of types (a) and (b).

4.7.3 JOINT DESIGN

Test information on all of the above types is very limited if it exists at all. The available data are summarized below.

Type (a) *Bearing properties.* Weaver[25] has tested bearing properties in several lay-ups of CFRP; 0°, 0°–90° and 'isotropic'. He shows that for edge distance/bolt diameter, $e/D > 4.0$ that the allowable bearing stress decreases

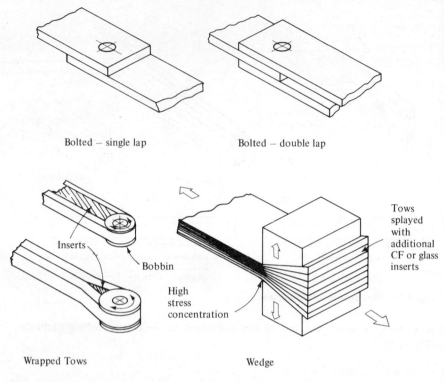

Bolted – single lap   Bolted – double lap

Inserts

Bobbin

High
stress
concentration

Tows
splayed
with
additional
CF or glass
inserts

Wrapped Tows   Wedge

Fig. 4.26 *Mechanical joints.*

with increasing diameter/thickness, D/t (Figs. 4.28, 4.29, and 4.30). Dastin[32] reports similar behaviour for a 181 style glass fabric. Althof and Muller's results[26] for CFRP are, however, at variance with the above but are in the same range (Fig. 4.28). Althof also quotes a 'proof' bearing stress based on a strain of 0·5 per cent $= (\Delta D/D)$ 100 per cent but Weaver's stress–strain curves are linear almost to failure for the equivalent joint (Fig. 4.31). Further data on CFRP[27,28] cannot be compared with the above as it is for specific lay-ups, but the values attained are 25 to 100 per cent higher. Comparable data for GFRP can be seen in references 29, 30, and 31.

Weaver also tested a limited number of simple lap joints using either bolts or rivets (Fig. 4.32) where failure occurred in two distinct stages; stress concentration precipitated an early failure at one bolt followed by final failure at the other.

Clifton[29] shows that the bearing stresses can be doubled in GFRP joints by torque tightening the bolts to a predetermined amount; unfortunately this cannot be relied upon for design purposes as stress relaxation in the matrix could result in a loss of preload.

136

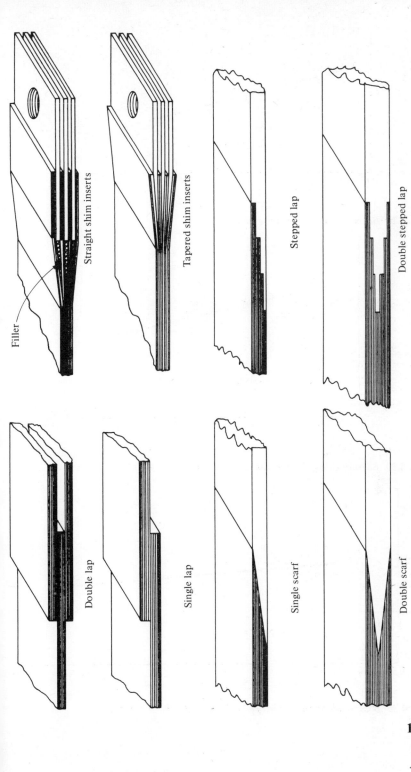

Filler

Straight shim inserts

Tapered shim inserts

Stepped lap

Double stepped lap

Double lap

Single lap

Single scarf

Double scarf

Fig. 4.27 *Bonded joints.*

137

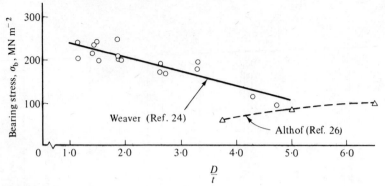

Fig. 4.28 *Bearing strength—unidirectional CF.*

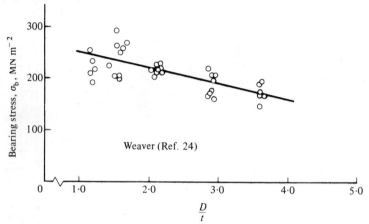

Fig. 4.29 *Bearing strength—crossply CF ($R = 0.5$, $\theta = 90°$).*

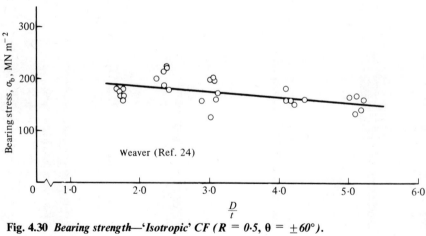

Fig. 4.30 *Bearing strength—'Isotropic' CF ($R = 0.5$, $\theta = \pm 60°$).*

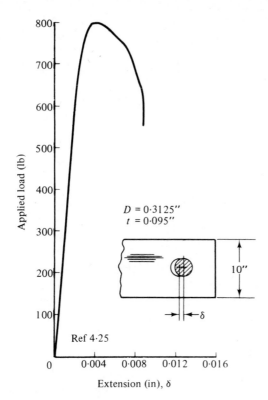

Fig. 4.31 *Load-extension UD specimen.*

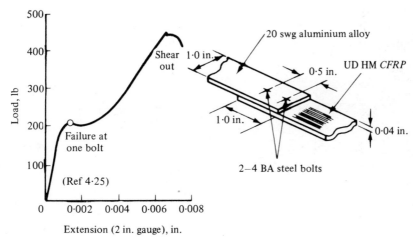

Fig. 4.32 *Load–extension, bolted lap joint.*

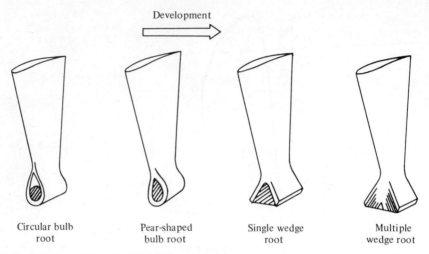

Circular bulb root   Pear-shaped bulb root   Single wedge root   Multiple wedge root

Fig. 4.33 *Rolls-Royce fan blade root development.*

It can be seen that this type of joint will have a higher weight than a metal joint as the bearing stresses are very low—approximately one-third of the values for a high strength aluminium alloy.

Type (a) Wedge joints. Joints have been made in CFRP using this method and the stages of development of a typical joint can be seen in Fig. 4.33, this being for a compressor blade root. The most successful was the multiple wedge[33] which has a load-carrying capacity up to three times that of the single wedge. The blade is of course fitted into a wedge-shaped slot. The first type tested consisted of continuous filaments wrapped around a pin but was unsuccessful due to the bending of the filaments and the damage sustained during subsequent hot press curing in matched moulds. A suggested adaptation of this joint for sandwich structures is shown in Fig. 4.34. Care must be

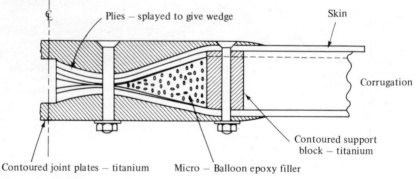

Plies – splayed to give wedge   Skin

Corrugation

Contoured support block – titanium

Contoured joint plates – titanium   Micro – Balloon epoxy filler

Fig. 4.34 *Wedge-type skin–stringer joint.*

140

exercised to support the filaments adequately where 'kink' loads are present. As the number of variables in these joints is high and the few tested are for a specific purpose (e.g., blade roots) data are not available on their comparative performance. Lehman and Palmer have designed and tested somewhat similar joints for a boron–epoxy strut[34] where the weight penalty incurred offset the weight reduction gained by use of the composite. The design assumptions were conservative and the results should not be taken too seriously.

Type (a) Wrapped joints. As commented above Rolls Royce had little success with this type but Dowty Rotol have used a joint for a compressor blade using filaments wrapped around a circular pin, the whole being placed in a wedge slot. No data are available. The German GFRP sailplane manufacturers have all used this type of joint without the wedge, transferring load to the pin in highly stressed areas very successfully but again the test data are not readily available. Bolköw have a similar joint for the root of their helicopter blade.[35]

Type (b) Bonded lap joints. The stress distribution in the adhesive between the two plates is similar to that in a metal–metal joint where

$$\frac{\tau \text{ average}}{\tau \text{ maximum}} = C . j . \tanh\left(\frac{l}{C . j}\right)$$

$$C = \left(\frac{2 . h . E_{\text{plate}}}{G_{\text{adhesive}}}\right)^{1/2}$$

$$j = \sqrt{(t)}/l$$

h = glue line thickness

t = plate thickness

Higher average shear stresses may be anticipated for high stiffness unidirectional composites for a given glue line thickness assuming that the bond is comparable at the fibre interface. The attained shear stresses will of course reduce considerably with simple lap joints in multi-directional composites. The trend can be seen for angle ply GFRP in Fig. 4.35 and the behaviour of CFRP will be similar. For simple angle ply the ratio of joint/plate strength will increase as the ply angle θ increases. Bonded lap joints in boron epoxy also exhibit similar characteristics.[36] Some results for a CFRP–aluminium alloy joint, with additional film adhesive, are shown in Fig. 4.36.

It is felt that this type of joint is not in itself suitable for joining heavy load-carrying members but it will be required for shear transfer over long lengths, for example flange-web connections.

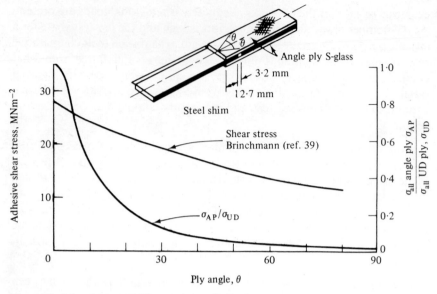

Fig. 4.35 *Joint shear strength v angle ply; S glass.*

Type (b) Bonded scarf joints. Some success has been achieved with both single[36,27] and double scarf joints[36] for boron and carbon composites. If the scarf angle can be accurately made and be of sufficiently shallow angle (say 3°) then joint strengths approaching that of the basic composite are possible. However, scarf angles of only 3° are not practical and 5° is suggested

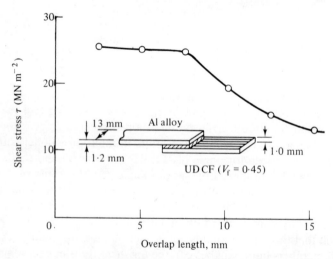

Fig. 4.36 *Aluminium alloy–UDCFRP lap joints.*[25]

142

as a minimum[37] although this will depend on the available adhesive shear strength. These joints should also have improved fatigue properties due to the reduced stress concentrations.

A more practical approach is a stepped scarf; eliminating the need for machining. Some joints have been made in boron–epoxy using this method with very low weight penalties compared with a metal joint.[34,37,38] This type of joint appears to be very attractive indeed for CFRP, and is worthy of development.

Type (b) Interleaved shims. Weaver[25] tested both stainless steel and titanium inserts in 0° CF composite achieving up to 27·6 MN m^{-2} for very short insert lengths. The most successful joints were those using an additional film adhesive between the metal and prepreg. The results for this type of joint are shown in Fig. 4.37. Brinchmann[39] has tested similar joints in various lay-ups of UD-GFRP but for only one insert length (12·7 mm); the results are shown in Fig. 4.35. He also comments on the difficulty of drilling holes through a stack of very thin shims (0·25 mm) and as a result limited shim thickness to 0·5 mm. This problem could be alleviated to some extent by

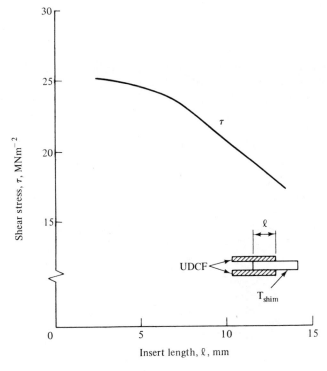

Fig. 4.37 *Shim joint—titanium–UDCF.*[25]

143

predrilling the shims, using these holes as tooling holes and reaming in situ after bonding.

It is apparent from the results that this type of joint can attain strengths comparable with the basic laminate with modest insert lengths (13 mm, say). Improvements in performance will also be possible with tapered shims but manufacture is likely to be more difficult.

Type (c) Combined bearing with interleaved shims. Wong[40] has tested and proposed a design method for this type of joint. The tests, however, are only for GFRP. The results are encouraging. Lehman[37] also tested similar joints in boron–epoxy with stainless steel inserts with very high failing loads. Few data are available for CFRP although Varlas[41] shows an improvement in bearing strength using interleaved asbestos paper. In view of the number of variables considerable effort is required in this area before any conclusions may be drawn.

4.8 Fatigue

Data are very limited and confined mainly to unnotched, unidirectional or crossply 0–90° specimens. Owen[42] has shown that for a unidirectional CF specimen the *S–N* curve is very flat, reducing to only 90 per cent of the short term tensile properties at 10^7 cycles. Harris and Beaumont[43] also confirm this and show that fibre treatments, resin system type, cyclic speed, wet or dry environment, etc., have little effect on the results. As can be expected, loading on axes other than the fibre axis drastically alters the picture, the behaviour then being nearer that of the resin system. Owen also shows that the fatigue strength is proportional to the number of axial plies in a 0–90° lay-up (Fig. 4.38).

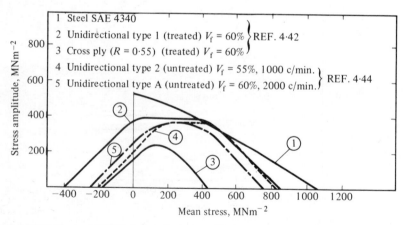

Fig. 4.38 *Fatigue of CFRP (unnotched) for 10^7 cycles.*

144

Colclough and Russell[44] have obtained very similar results for UD-CFRP in their propeller blade test programme and the comparison can be seen in Fig. 4.38.

If these values of fatigue strength were expressed in terms of specific fatigue strength (i.e., permissible stress/density) then for UD-CFRP this may be seen to be up to five times that of HS steel and up to three times that of aluminium alloy. Like the simple merit index this can of course be very misleading but it does give some indication of the potential of the material.

In a complex load situation CF would be difficult to justify in terms of fatigue behaviour only, as holes, notches, etc., would seriously degrade its properties. Nevertheless in those areas where fatigue is only one of many requirements it is still very competitive. For example the main design case may be compressive with the secondary case lower tensile loading.

The preliminary results obtained are encouraging and both Dowty Rotol[44] and Rolls-Royce[34] have succeeded in developing CF propeller and fan blades respectively with satisfactory fatigue characteristics. The relatively poor performance of CF for off-axis loading can be avoided by correct choice of lay-up for a given load combination.

4.9 Creep

Creep of UD-CFRP is of negligible proportions at normal working temperatures. Unlike woven fabric reinforcement, where straightening of the filaments under tension or compression induces high strains in the matrix, UD-CFRP has very low strain. If a multi-layer composite is correctly designed for a single load case the load carried by the matrix will be negligible and creep reduced to a minimum. If other loading cases of a different direct/shear load ratio are then applied, thus imposing higher loads on the matrix, creep may be significant. Random or short fibre, aligned mat will creep to a greater degree as the resin matrix plays a much more significant role. Information on creep of the latter and of UD-CFRP is very sparse.

Another problem area may be that of creep buckling where sustained compressive or shear loading is present. The inevitable eccentricities in the filaments or plates will induce loads in the matrix which would not be present in a perfect laminate, as a result creep buckling could occur. Parallel to this is the post-buckling behaviour of plates and struts. In some areas the ideal non-buckled design may be impractical (e.g., required stiffener pitch too small) and a design where part of the structure buckles at a fraction of maximum load must be used. In this case knowledge of post-buckling behaviour is vital. We can, however, conservatively assume that the element will continue to support the load at which it buckled. It is felt that this latter type of design is undesirable with CF structure as the problem of creep buckling

145

will be accentuated. Alternative structures can be designed where this problem is minimized, for example by use of sandwich construction.

4.10 Applications

4.10.1 AEROSPACE STRUCTURES

In structures where reduction in weight produces a direct trade-off in improved performance and/or a smaller vehicle there are many possible applications. The material will be introduced initially in the secondary structure where failure would not be catastrophic. As experience and confidence is gained CF will be utilized for the primary load-carrying structure where the maximum weight savings could be made.

4.10.2 AIRBUS PROJECT

The methods applied previously have been used for the main wing structure of an airbus project.[23] This item was chosen as it has a relatively simple load envelope and an absence of large cut-outs. Effort was concentrated on the torsion box (Fig. 4.39), this being essentially a long, shallow box resisting high bending and shear loads. The skins carry high compressive or tensile direct loads due to bending and relatively low shear loads. The webs resist load due to shear and torque. In addition there is a specified torsional stiffness criterion.

The original structure was designed in high strength aluminium alloy and was redesigned for HMCF, the latter being chosen for buckling and torsional stiffness considerations. The top skin was initially designed for compressive load only as the direct/shear load ratio is approximately ten and consequently the effect of shear on buckling is small. A rapid assessment of the required equivalent skin thickness for various compression panel geometries was possible by reference to Fig. 4.20, together with optimum lay-up (i.e., $R = 0.33$, $\theta = 55°$). The total equivalent skin thickness of skin and ribs was then determined for a range of rib pitches (or strut length) L. The webs were redesigned for CF using a $\pm 45°$ lay-up for maximum shear stiffness resulting in a weight penalty of 20 per cent compared with a $\pm 60°$ lay-up. The weight saving for the webs compared with the aluminium alloy using an increased joint weight penalty was 22 per cent.

The torsional stiffness of the above design was then checked and found to be lower than that required. By reference to Fig. 4.40 it can be seen that by altering the angle ply from the optimum $\pm 55°$ to $\pm 45°$, the allowable compressive stress remains constant, the shear stiffness is increased by 11 per cent accompanied by a negligible loss in buckling efficiency, Z, of 1 per cent. The skins were then redesigned, using the modified lay-up and also by adding sufficient $\pm 45°$ layers to the surface, to match the torsional stiffness of the

146

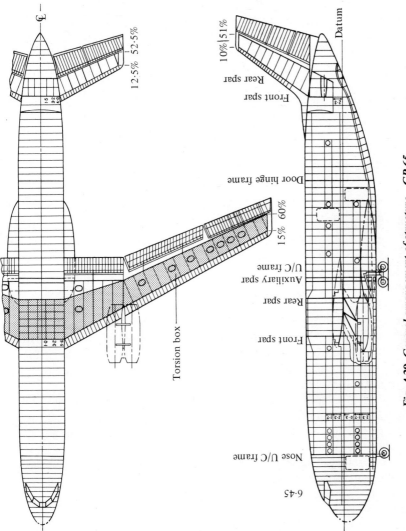

Fig. 4.39 *General arrangement of structure—GP 65.*

147

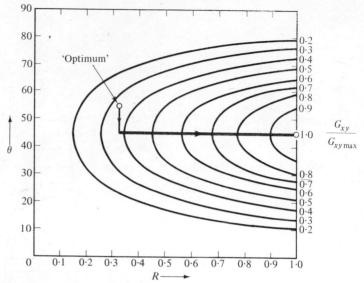

Fig. 4.40 *Variation of shear modulus.*

original aluminium alloy wing. Allowing for joint weights the overall weight saving was 57 per cent.

The design of fuselage structures with varying loading envelopes is complicated by the presence of large cut-outs for windows and doors. However, even using a simple pseudo-isotropic lay-up (i.e., $R = 0.67$, $\theta = \pm 60°$) produces a material with axial and shear strength and stiffness comparable with high strength aluminium alloy and slightly lower transverse strength yet only two-thirds as heavy. On this simple basis weight savings of 30 per cent appear possible and in most situations the material lay-up could be favourably oriented.

In the more lightly loaded structure the possible weight savings will be reduced as minimum thickness limits, based on handling or accidental damage criteria, are reached. These minimum thicknesses are yet to be specified. Fortunately these areas are of relatively low weight.

As laboratory material properties were used for the study and no additional 'plastic' safety factors used the above merely shows the potential weight saving. Improvements in material performance and predictability could well enable this potential to be realized. The resulting improved vehicle performance will justify additional raw material costs and there is considerably less material by weight.

4.10.3 STRIKE AIRCRAFT WING

Dickinson[45] has predicted weight savings of 30 per cent using structure similar to that for the airbus wing. For this structure the torsional stiffness requirement is more severe.

4.10.4 SUPERSONIC AERIAL TARGET

McQuillen and Huang[46] have recently designed a wing for a target vehicle using very similar methods and have confirmed their predicted weight savings of 50 per cent by laboratory specimen tests.

4.10.5 V/STOL AIRCRAFT STRUCTURE

Dukes and Krivetsky[47] predict overall weight savings of 30 per cent by extensive use of CF.

4.10.6 GLIDER WING SPARS

The wings of many modern, high aspect ratio (i.e., span/chord ratio) sailplanes are built almost wholly of GFRP, the material being used very effectively. The object is to produce an easily built, very smooth surfaced wing. Metal glider wing skins tend to buckle due to wing bending unless supported by close-pitch stiffeners or by use of sandwich structure. The GFRP construction avoids this problem by using high spar/skin stiffness ratios and thus carrying the major part of the bending loads in the spar caps. This is achieved by using spanwise unidirectional glass rovings for the spar caps (Fig. 4.25) and woven GFRP at $\pm 45°$ for the skins and webs. Accurate wing profiles are maintained by use of sandwich skins, the latter being laid up by hand in female moulds. The skins and webs are also favourably orientated to provide maximum torsional stiffness. This is not a new method of construction; wooden gliders are similar, but GFRP has high strength and lends itself to small-scale production using semi-skilled labour. It would appear that this is an ideal solution to a difficult problem. Unfortunately the search for increased performance has led to machines with high aspect ratios and long spars (up to 23 m) and since these wings are alarmingly flexible in bending the design criterion is now flexural stiffness, not strength. A metal spar would reduce the problem but is less compatible with a GFRP skin and the method of manufacture (i.e., small-scale plant). An obvious alternative is to use UD-CFRP for the spar cap using virtually identical production methods to those used for glass. For the same weight spar cap the wing would be approximately five times as stiff in flexure and probably have an improved fatigue life. Alternatively the aspect ratios could be considerably increased. Slingsby Sailplanes[48] have already built a prototype high performance (19 metre span) sailplane using CF in place of the GF spar. The result is a sailplane with a very stiff wing, and a spar cap which cost in itself twice the cost of two complete GFRP machines!

4.10.7 PRESSURE VESSELS

These have already been made by several firms in CF using equipment formerly used for GFRP filament winding. The advantages to be gained in CF are that its high specific stiffness results in considerably less strain in the resin matrix. As a result, unlike GFRP 'weeping' does not occur until a high percentage of ultimate strength is reached. The reduced dilation of the vessels also facilitates the insertion of metal joining rings where open ended vessels are necessary (e.g., cartridge type rocket motors). Where cyclic loading is possible then CF has a marked advantage over glass due to its better fatigue properties. Optimum orientations can be determined by the methods already proposed.

4.10.8 COMPRESSOR AND PROPELLER BLADES

These have been successfully designed in CF,[33,44] the weight of the propeller blades being approximately 60 per cent of the aluminium equivalent. The reduction in centrifugal force in high speed rotating machinery due to decreased component weight plays a significant part in the achieved weight reductions. A major problem with these components is erosion and in the case of the Dowty Rotol compressor blade the protection has taken the form of polyurethane sheathing, whilst the RR fan blade had a leading edge of very thin titanium. Bird impact is also a problem, particularly with compressor blading and it has been suggested that three-dimensional weaving (inserting UD-CF perpendicular to the plane of the plates) improves the energy absorption capacity. Unfortunately it also degrades the specific strength and stiffness. The bird strike problem is by no means solved at this time.

4.10.9 CF REINFORCED METAL STRUCTURES

BAC[49] have paid particular attention to the design of metal beams with the beam caps stiffened with precured strips of UD-CF (Fig. 4.41). Overall

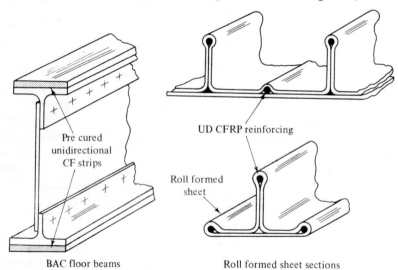

BAC floor beams   Roll formed sheet sections

Fig 4.41 *CF reinforced metal structures.*

150

weight savings of 2 to 3 lb for each 1 lb of CF used are suggested.[50] As it is now possible to produce integrally formed structures from sheet metal (Fig. 4.41) with hollow sections, this would appear to be ideally suited for the insertion of UD-CF. As the coefficient of expansion of UD-CF is very low (see appendix) variations in temperature could induce high thermal stresses at the metal–composite interface. This is particularly difficult if metal panels are reinforced with CF stiffeners. Increases in temperature could cause local buckling of the metal and probably severe bowing of the panels. The weight savings with this type of construction are unlikely to be as great as with those constructed wholly in CF.

4.10.10 CF REINFORCED GFRP

This has already been referred to in the section on the glider wings. Other possibilities are the use of sandwich construction with CF face sheets and GFRP core, giving panels with high flexural stiffness and improved shear stiffness. The techniques for design of this type of structure are identical to those already described for those made wholly from CF and those with an additional reinforcing grid of strips of CF on one side of a GFRP structure. The latter is an inefficient use of CF but it does result in a stiffer structure for a given weight and has some use for existing contact moulded structures. To facilitate this method UD-CF tows are now available on a woven glass backing tape. The kinking of the tows at the grid node points does of course seriously limit its use for high strength applications.

The latter method has been used for stiffening thin shell GFRP structures such as racing dinghies, racing and sports car bodies and high speed racing boats. A study of the optimum proportion of CF to GFRP or metal and the preferred positioning of CF in structural panels is being undertaken at present.[51]

4.10.11 MISCELLANEOUS STRUCTURES

Possible weight savings for various structural components are shown below.

Table 4.3 *Percentage weight saving compared with high strength aluminium alloy*

| Component | Loading Index Level | | Reference |
	Low	High	
Wide Columns	43%	60% (optimum)	
(Plate Stiffener)		73% (unidirectional)	Fig. 4.20
Plates (Compression)	42%	73%	Fig. 4.17
Plates (Shear)	42%	62%	Fig. 4.24
Corrugated Plates (Shear)	8%	62%	Fig. 4.24

The above predictions are verified by Dukes and Krivetsky[47] for similar components.

Hieronymus[52] quotes weight savings of 30 per cent, 23 per cent, and 20 per cent for a fighter fin, speed brake, and horizontal stabilizer respectively.

The weight savings for heavily loaded torque tubes are shown in table 4.1. Winny[28] has determined weight savings for high speed transmission shafting of approximately 50 per cent with aluminium alloy.

Although little attention has been paid to random CF mat it has in fact several possible applications. Its short-term specific modulus is the equivalent of steel and aluminium alloy and as such is particularly attractive for complex double curvature components such as gearbox casings, thin shell structures, etc. Design methods would of course be similar to those used for conventional metal structures, providing that time dependence of the material was catered for. The material would thus be very useful for the manufacture of prototypes using hand lay-up methods and if material costs reduce perhaps even for low production runs.

4.11 Problem Areas

The question of adequate quality control of the completed structure is discussed in detail in chapter 7. As with all bonded structures the production of full-scale components having properties equivalent to those achieved in laboratory test specimens can only be by carefully controlled manufacture, probably in 'clean room' conditions. Even the use of coupon specimens made simultaneously with the component does not always ensure the desired quality. Further development of suitable non-destructive testing methods is vital (see chapter 7).

Safety factors are closely related to quality control and current UK practice for aircraft structures[53] is to use an additional safety factor or 'superplastic' factor of 1·5. American practice on the other hand is to leave the choice of a 'superplastic' factor, if any, to the discretion of the designer based on the results of full-scale or coupon testing. As with metal structures these factors will be combined with statistically derived design allowable stresses such that not more than a certain percentage of specimens fail below a quoted value. These stress levels have not yet been satisfactorily determined even for UD specimens and certainly not for those with angle plies. Preliminary work by Taylor[54] suggests that the designed elastic moduli can be achieved with reasonable accuracy as the effects of flaws, voids, etc., are not so damaging. There is, however, considerably more scatter shown by most experimenters for composite strengths. It would seem reasonable therefore to use two separate factors depending upon whether the component is designed on strength or stiffness. In the case of a compression or shear panel this can be determined by reference to Figs. 4.16, 4.20, 4.22 and the appropriate factors included as shown in Fig. 4.42. With 'superplastic' factors as high as 1·5 considerable weight savings are still possible.[28]

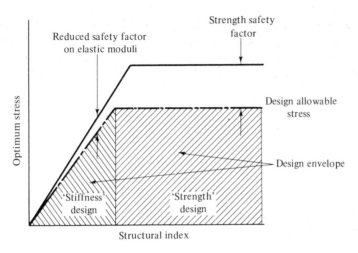

Fig. 4.42 *Application of safety factors.*

Many structures, aircraft, tall masts, etc., must be protected from the possibility of damage due to lightning strike. In the case of metal structures this is achieved by electrical bonding together of all components thus providing a path to conduct the sudden surge of electric current. In the case of non-metallic structures a lightning conductor is added (glass fibre fuel tanks have a copper grid bonded on the surface). Some concern has been expressed regarding the safety of adhesive bonded metal structures where a strike can cause local damage and more seriously in the case of metal honeycomb sandwich 'explosive' debonding due to vaporization of the honeycomb foil.[55] Kelly[56] suggests that protective measures in the form of conducting strips bonded to the exterior surface plus a silver-particle-filled, epoxy paint finish or a thin, flame-sprayed, aluminium coating may offer a partial solution. Fassell[57] has shown that a graphite (Thornel) epoxy composite can withstand high current densities for very short periods without apparent degradation. Fray[58] dismisses this problem based on limited testing and assumes, possibly correctly, that it can be overcome by protective sheathing with a very small weight penalty.

Electrolytic corrosion caused by potential differences between materials in contact, for example between carbon and aluminium or steel, may well be delayed if not eliminated by the presence of the insulating adhesive film. Some short-term tests have been performed[58] with satisfactory results but long-term data are required including environmental effects (see p. 185).

The very low coefficient of thermal expansion could result in high residual stresses if CF were hot cured on to metal substructure although they will reduce with time due to stress relaxation in the matrix interface. This can be overcome by cold curing the precured reinforcing elements on to the metal

153

thus eliminating part of the problem. Subsequent temperature variations could give rise to thermal stress but in most areas this is likely to be less of a problem. It is suggested that wherever possible all-CF components be designed rather than mixed CF-metal but the latter should not be ignored.

4.12 Conclusion

In spite of the problems just discussed and the difficulty of designing for variations in the loading pattern, CF composites can be used to advantage particularly in those areas where weight saving has priority. Considerable time and effort is required to determine allowable stress levels, approved manufacturing methods including quality control, etc., before primary structures may be confidently designed. The potential advantages have been demonstrated and the way lies open for a big step forward in the manufacture of more efficient lightweight structures.

4.13 Sample Strength Calculation

4.13.1 SINGLE LAYER PROPERTIES

$$E_\alpha = 207 \cdot 6 \times 10^3 \text{ MN m}^{-2}; \qquad E_\beta = 7 \cdot 59 \times 10^3 \text{ MN m}^{-2}$$

$$G_{\alpha\beta} = 4 \cdot 83 \times 10^{-3} \text{ MN m}^{-2}; \qquad \mu_{\alpha\beta} = 0 \cdot 3$$

$$X_{\text{comp}} = 827 \cdot 6 \text{ MN m}^{-2}; \qquad Y = 41 \cdot 4 \text{ MN m}^{-2}; \qquad S = 55 \cdot 2 \text{ MN m}^{-2}.$$

4.13.2 MULTI-LAYER STIFFNESS

Consider a three-ply system where $R = 0 \cdot 4$, $\theta = \pm 60°$, plies evenly distributed.

$$b_{ij\,\text{comp}} = 10^3 \begin{bmatrix} 0 \cdot 6 \times 207 \cdot 6 + 0 \cdot 4 \times 21 \cdot 7, \, 0 \cdot 6 \times 2 \cdot 28 + 0 \cdot 4 \times 38 \cdot 15, 0 \\ 0 \cdot 6 \times 2 \cdot 28 + 0 \cdot 4 \times 38 \cdot 15, 0 \cdot 6 \times 7 \cdot 6 + 0 \cdot 4 \times 121 \cdot 7 \,, 0 \\ 0 \qquad\qquad\qquad\qquad , 0 \qquad\qquad\qquad , 0 \cdot 6 \times 4 \cdot 8 + 0 \cdot 4 \times 40 \cdot 7 \end{bmatrix}$$

therefore

$$b_{ij\,\text{comp}} = 10^3 \begin{bmatrix} 133 \cdot 3 & 16 \cdot 6 & 0 \\ 16 \cdot 6 & 53 \cdot 2 & 0 \\ 0 & 0 & 19 \cdot 2 \end{bmatrix} ; \qquad a_{ij\,\text{comp}} = [b_{ij}]^{-1}$$

therefore

$$a_{ij\,\text{comp}} = \begin{bmatrix} 7 \cdot 80 & -2 \cdot 44 & 0 \\ -2 \cdot 44 & 19 \cdot 54 & 0 \\ 0 & 0 & 52 \cdot 15 \end{bmatrix} \times 10^{-6}$$

154

therefore

$$E_x = \frac{1}{7 \cdot 80} \times 10^6 = 128 \cdot 2 \times 10^3 \text{ MN m}^{-2}$$

$$E_y = \frac{1}{19 \cdot 54} \times 10^6 = 51 \cdot 2 \times 10^3 \text{ MN m}^{-2}$$

$$G_{xy} = \frac{1}{52 \cdot 15} \times 10^6 = 19 \cdot 2 \times 10^3 \text{ MN m}^{-2}$$

$$\mu_{xy} = \frac{2 \cdot 44}{7 \cdot 80} = 0 \cdot 313$$

$$\mu_{yx} = \frac{2 \cdot 44}{19 \cdot 54} = 0 \cdot 125$$

4.13.3 MULTI-LAYER STRENGTH

This must be calculated for a specified ratio of applied stresses e.g., $\sigma_x = -10$ MN m^{-2}, $\sigma_y = 0$, $\tau_{xy} = 0$. Now

$$
\begin{bmatrix} \varepsilon_x \\ \varepsilon_y \\ \gamma_{xy} \end{bmatrix} = [a_{ij}] \begin{bmatrix} \sigma_x \\ \sigma_y \\ \tau_{xy} \end{bmatrix} = \begin{bmatrix} 7 \cdot 80 & -2 \cdot 44 & 0 \\ -2 \cdot 44 & 19 \cdot 54 & 0 \\ 0 & 0 & 52 \cdot 15 \end{bmatrix} \times \begin{bmatrix} -10 \\ 0 \\ 0 \end{bmatrix} \times 10^{-6}
$$

therefore

$$
\begin{bmatrix} \varepsilon_x \\ \varepsilon_y \\ \gamma_{xy} \end{bmatrix} = \begin{bmatrix} -0 \cdot 07800 \\ 0 \cdot 02436 \\ 0 \end{bmatrix} \times 10^{-3}
$$

Consider now stresses in each layer.

For Layer $\theta = 0°$

$$
\begin{bmatrix} \sigma_{x0} \\ \sigma_{y0} \\ \tau_{xy0} \end{bmatrix} = [b_{ij0}] \begin{bmatrix} \varepsilon_x \\ \varepsilon_y \\ \gamma_{xy} \end{bmatrix} = \begin{bmatrix} 207 \cdot 6 & 2 \cdot 28 & 0 \\ 2 \cdot 28 & 7 \cdot 61 & 0 \\ 0 & 0 & 4 \cdot 83 \end{bmatrix} \begin{bmatrix} -0 \cdot 0780 \\ 0 \cdot 02436 \\ 0 \end{bmatrix}
$$

therefore

$$
\begin{bmatrix} \sigma_{x0} \\ \sigma_{y0} \\ \tau_{xy0} \end{bmatrix} = \begin{bmatrix} -16 \cdot 14 \\ 0 \cdot 0075 \\ 0 \end{bmatrix} \text{ MN m}^{-2}
$$

For Layer at $\theta = +60°$

$$\begin{bmatrix} \sigma_{xT} \\ \sigma_{yT} \\ \tau_{xyT} \end{bmatrix} = \begin{bmatrix} 21.7 & 38.15 & -22.6 \\ 38.15 & 121.7 & -64.0 \\ -22.6 & -64.0 & 40.7 \end{bmatrix} \begin{bmatrix} -0.0780 \\ 0.02436 \\ 0 \end{bmatrix} = \begin{bmatrix} -0.76 \\ -0.011 \\ 0.201 \end{bmatrix}$$

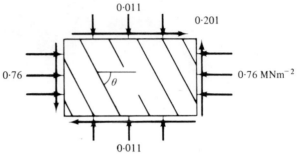

The stresses in the θ layer must now be transferred to the filamentary axes α, β.

$$\text{therefore} \quad \begin{bmatrix} \sigma_{\alpha T} \\ \sigma_{\beta T} \\ \tau_{\alpha \beta T} \end{bmatrix} = [mn] \begin{bmatrix} \sigma_{xT} \\ \sigma_{yT} \\ \tau_{xyT} \end{bmatrix}$$

$$\text{where } [mn] = \begin{bmatrix} \cos^2 \theta & \sin^2 \theta & -2 \sin \theta \cos \theta \\ \sin^2 \theta & \cos^2 \theta & 2 \sin \theta \cos \theta \\ \sin \theta \cos \theta & -\sin \theta \cos \theta & (\cos^2 \theta - \sin^2 \theta) \end{bmatrix}$$

Then for $+\theta$ layer

$$\begin{bmatrix} \sigma_{\alpha T} \\ \sigma_{\beta T} \\ \tau_{\alpha \beta T} \end{bmatrix} = \begin{bmatrix} 0.25 & 0.75 & -0.866 \\ 0.75 & 0.25 & 0.866 \\ 0.433 & -0.433 & -0.50 \end{bmatrix} \begin{bmatrix} -0.76 \\ -0.011 \\ 0.201 \end{bmatrix} = \begin{bmatrix} -0.372 \\ -0.399 \\ -0.425 \end{bmatrix}$$

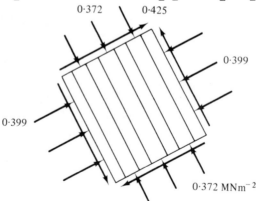

NOTE

For $\theta = -60°$ layer the procedure is similar to that for $\theta = 60°$ but b_{13}, b_{23} are negative.

The stresses in each layer are then compared with the chosen failure criteria, e.g. Hill Yield Criteria where

$$K = \frac{X}{\sqrt{[\sigma_\alpha^2 - \sigma_\alpha \cdot \sigma_\beta + (X^2/Y^2) \cdot \sigma_\beta^2 + (X^2/S^2) \cdot \tau_{\gamma\alpha}^2]}}$$

and

K = factor of safety

X, Y, S are measured strengths

$\sigma_\alpha, \sigma_\beta, \tau_{\alpha\beta}$ are stresses referred to filamentary axes

Then for $\theta = 0°$

$$X = 827\cdot6 \text{ MN m}^{-2} \text{ (compression)}$$

$$Y = 41\cdot4 \text{ MN m}^{-2}$$

$$S = 55\cdot2 \text{ MN m}^{-2}$$

therefore

$$K_0 = \frac{827\cdot6}{\sqrt{[(16\cdot14)^2 + 16\cdot14 \times 0\cdot0075 + (827\cdot6/41\cdot4)^2 \cdot 0\cdot0075^2]}}$$

$$= \underline{51\cdot27}$$

Similarly for $\theta = 60°$

$$K_T = \frac{827\cdot6}{\sqrt{[(0\cdot372)^2 - 0\cdot372 \times 0\cdot399 + (827\cdot6/41\cdot4)^2 \cdot (0\cdot399)^2 + (827\cdot6/55\cdot2)^2 \cdot(0\cdot425)^2]}}$$

therefore

$$K_T = \underline{81\cdot05}$$

That is, primary failure occurs in $0°$ layer at an applied stress level, $\sigma_x = -10 \times 51\cdot27 = -512\cdot7 \text{ MN m}^{-2}$.

References

1 Cox, H. L., 'The elasticity and strength of paper and other fibrous materials', *British Journal of Applied Physics*, 18 May 1951.
2 ROTHWELL, A., 'Optimum fibre orientation for the buckling of thin plates of composite material', *Fibre Science and Technology*, Oct. 1969.
3 SHIBLEY, A. M., 'Filament Winding', *Handbook of Fibreglass and Advanced Plastic Composites*, ed. G. Lubin, Von Nostrand, 1969.

4 HARRIS, G. Z., *Optimum Fibre Arrangements for Reinforced Sheets under Combined Loading*, RAE, TN STRUCTURES TR66361,1968.
5 TSAI, S., *Structural Behaviour of Composite Materials*, NASA CR-71,1964.
6 ROSEN, B. W., DOW, N. F., and HASHIN, Z., *Mechanical Properties of Fibrous Composites*, NASA CR-31, April 1964.
7 CHAMIS, C. C., *Failure Criteria for Filamentary Composites*, NASA TN D-5367, August 1969.
8 TSAI, S., *Fundamental Aspects of F.R.P. Composites*, chapter 1, ed. R. T. and H. S. Schwartz, Interscience, 1968.
9 ANON, *Plastics for Aircraft: Pt. 1 Reinforced Plastics*, ANC-17, June 1955.
10 HILL, R., *The Mathematical Theory of Plasticity*, Clarendon Press, 1950.
11 TSAI, S., *Strength Characteristics of Composite Materials*, NASA CR-224, April 1965.
12 TIMOSHENKO, S. P. and GERE, J. M., *Theory of Elastic Stability*, McGraw-Hill, 1961.
13 HOFF, N. J., *Engineering Laminate*, chapter 1, ed. A. G. H. Dietz, John Wiley and Sons, 1949.
14 ANON, *Wooden Aircraft Structures*, ANC-18.
15 DAVIS, J. G., and ZENDER, G. W., *Compressive Behaviour of Plates Fabricated from Glass Filaments and Epoxy Resin*, NASA TN D-3918, April 1967.
16 HARRIS, C. M., *Buckling of F.R.P. Plates*, Cranfield Institute of Technology Thesis, September 1968.
17 BRAIN, C. J., *Buckling of Filament Reinforced Plastic Plates*, Cranfield Institute of Technology Thesis, September 1969.
18 WITTRICK, W. H., 'Correlation between some stability problems for orthotropic and isotropic plates under bi-axial and uni-axial direct stress', *Aero. Quarterly*, 4, 1952–1954.
19 ANON, Royal Aeronautical Society Data Sheets—*Structures*, Vol. 1, DS.02.01.01.
20 GERARD, G., *Minimum Weight Design of Compression Structures*, John Wiley and Sons.
21 EMERO, D. H., and SPUNT, L., 'Optimisation of Multirib and Multiweb Wing Box Structures under Shear and Moment Loads', A.I.A.A. 6th Structures and Materials Conference, Palm Springs, Calif., April 1965.
22 RICHARDS, D. M., *Optimum Design of Fibre Reinforced Corrugated Compression Panels*, College of Aeronautics Report AERO. No. 209, January 1969.
23 TETLOW, R., *Application of Carbon Fibre Composites to an Airbus*, Cranfield Institute of Technology Memo No. 10, 1971.
24 TETLOW, R., *Design Charts for Carbon Fibre Composites*, Cranfield Institute of Technology Memo No. 9, 1970.
25 WEAVER, C., *Joints in Carbon Fibre Reinforced Plastic*, Cranfield Institute of Technology Thesis, September 1970.
26 ALTHOF, W., and MULLER, J., 'Untersuchungen an Geklebten und Losbaren Verbindungen von Faserverstarkten Kunststoffen'. *Kunststoffe*, 60, 1970.
27 ZABORA, R. F., and HOGGATT, J. T., 'Joint Concepts for Carbon Composite Structures', *SAMPE PROC. National Technical Conf.*, Seattle, September 1969.
28 WINNY, H. F., 'The Use of Carbon Fibre Composites in Helicopters', *Aero. Journal*, December 1971.
29 CLIFTON, F. W. G., *Preliminary Experiments on Tensile Loading of Plain and Jointed Fibre-Reinforced Material*, RAE Structures Dept. Tech. Note 1708.
30 DALLAS, R. N., 'Mechanical Joints in Structural Composites', *12th National SAMPE Symposium*.
31 BRIEDENBACH, L., 'Composite Materials for Lift, Structure, Shape and Service in Deep Submersibles', *12th National SAMPE Symposium*.
32 DASTIN, S., *Handbook of Fibreglass and Advanced Plastics Composites*, ed. G. Lubin, Van Nostrand, 1969.
33 GRESHAM, H. E., *The Development of Fibre Reinforced Composites for Gas Turbines*, Inst. of Prod. Engineers, May 1969.
34 LEHMAN, G. M., and PALMER, R. J., 'Design and Development Study of Aircraft Structural Composites', *12th National SAMPE Symposium*.
35 ANON, 'Rigid Hub with Flexible Blades Extends Rotor Life', *Design and Components in Engineering*, 29 Jan. 1968.
36 ROGERS, C. W., *Fundamental Aspects of Fibre Reinforced Plastic Composites*, ed. H. S. Schwartz, Interscience, 1968.

158

37 LEHMAN, G. M., 'Fundamentals of Joint Design for Composite Airframes', Westec Conf. Los Angeles, March 1969.
38 OKEN, S., and JUNE, R. R., *Metal Structures Reinforced with Filamentary Composites*, NASA CR-1859.
39 BRINCHMANN, A., 'Segmented Fiberglass Motor Case Joint Design and Analysis', *Modern Plastics*, December 1966.
40 WONG, J. P., COLE, B. W., and COURTNEY, A. L., 'Development of the Shim Joint Concept for Composite Structural Members', *J. Aircraft* **6**, No. 1, Jan.–Feb. 1969.
41 VARLAS, M., 'Development of High Modulus Graphite/Asbestos Reinforced Laminates', *SAMPE Nat. Tech. Conf.*, Seattle, September 1969.
42 OWEN, M. J., 'How Fatigue Affects Glass and Carbon Fibre Composites', *Original Equipment and Design*, January 1972.
43 BEAUMONT, P. W. R., and HARRIS, S., *International Conference on Carbon Fibres*, Paper 49, The Plastics Institute, London, 1971.
44 COLCLOUGH, W. J., and RUSSELL, J. G., 'The Development of a Composite Propeller Blade with a Carbon Fibre Reinforced Plastics Spar', *Aero. Journal*, January 1972.
45 DICKINSON, J., *The Improvements in Performance from using Carbon Fibre Composite in the Design of a Strike Aircraft Wing*, Cranfield Institute of Technology Thesis, September 1971.
46 McQUILLEN, E. J., and HUANG, S. H., 'Graphite-Expoxy Wing for BQM-34E Supersonic Aerial Target', *J. Aircraft*, **8**, No. 6, June 1971.
47 DUKES, W. H., and KRIVETSKY, A., *The Application of Graphite Fibre Composites to Airframe Structures*, SAE 680316.
48 TUCKER, J., 'Sparring Partners', *Flight International*, 28 Oct., 1971.
49 McELHINNEY, D. M., KITCHENSIDE, A. W., and ROWLAND, K. A., 'The Use of Carbon Fibre Reinforced Plastics', *Aircraft Engineering*, October 1969.
50 SANDERS, R. C., 'The Effect of Carbon Fibre Composites on Design', *Aero. Journal*, December 1971.
51 TETLOW, R., *Design of F.R.P./Metal Structural Components*, C.I.T. (to be published).
52 HIERONYMOUS, W. S., 'Carbon Composites Program Gains', *Av. Week and Space Tech.*, August 18, 1969.
53 ANON, *Military Airworthiness Requirements*, Av. P. 970.
54 TAYLOR, C., *Failure Criteria for Filamentary Reinforced Plastics*, Cranfield Institute of Technology Thesis, September 1970.
55 ROBB, J. D., 'Mechanisms of Lightning Damage to Composite Materials', *Lightning and Static Electricity Conf.*, Pt. II. AFAL-TR-68-290, December 1968.
56 KELLY, L. G., and SCHWARTZ, H. S., 'Investigation of Lightning Strike Damage to Epoxy Laminates Reinforced with Boron and High Modulus Graphite Fibers', *Lightning and Static Electricity Conf.*, Pt. II. AFAL-TR-68-290, December 1968.
57 FASSELL, W. M., PENTON, A. P., and PLUMER, J. A., 'The Susceptibility of Advanced Filament Organic Matrix Composites to Damage by Simulated Lightning Strikes', *Lightning and Static Electricity Conf.*, Pt. II AFAL-TR-68-290, December 1968.
58 FRAY, J., 'A Carbon Fibre Vulcan Airbrake Flap', *Aero. Journal*, **75**, December 1971.

5. Mechanical Engineering Applications

M. Bedwell

Proposed uses for carbon fibre are legion. Apart from the aerospace applications for which the new material was explicitly developed, artefacts from cars to kites have been the objects of investigation.[1] But the engineer, overwhelmed by a surfeit of publicity, can be forgiven for asking 'which of these ideas are realistic for the near future—and how near is "near"?' It is the object of this chapter to answer these questions by reasoned argument from the properties carbon fibre has to offer. Such an approach necessarily leads to applications not normally considered under the heading of 'mechanical engineering', and in such cases detailed descriptions are left to the appropriate chapter.

The straight answer to the question 'what are carbon fibres being used for?' is 'in making test pieces'. To elaborate, the greater part of the output of the last four years has gone to the laboratories of high-technology industries for the purposes of material evaluation. Other industries wishing to use carbon fibre have been faced with a choice: either they can wait until they can profit from the spin-off of these evaluations, or, recognizing that in general engineering a greater ignorance of a material behaviour can be accepted, they can begin their own, shorter-term, developments. In each case there is an obstacle to writing about realistic uses of carbon fibre, for either these are still sound but untried ideas existing only on paper, or they are guarded with some jealousy by firms who understandably want to reap where they have had the commercial courage to sow. Discussion of applications therefore has to steer a middle course between speculation and respect for commercial secrecy.

To filter off the possible from the probable and actual, we will for the greater part of this chapter assume three basic premises: first, that carbon fibre is useful only in reinforcing other matrix materials, second, that the only established matrix is thermosetting resin, and third, that in volume or weight terms carbon fibre will always be much more expensive than the materials it will replace. In fact none of these premises is entirely justified, and will indeed be challenged in the concluding part of the chapter. But for the present they serve as logic filters to separate the speculative from the immediately probable.

The first and second premises mean that we dismiss immediately any suggestions where the conditions are clearly incompatible with plastics in any form, notably where the temperature is much above 250°C. The journalistic description of carbon fibre as 'ten times stronger than steel', while not entirely untrue, led many engineers who should have known better to presume that the novel reinforcement somehow also conferred on its matrix other steel-like qualities like refractoriness and hardness. This it most certainly does not, and the more bluntly and loudly this is said, the more effectively will the realistic applications enjoy the concentration of effort they deserve.

The third premise, that of expense, leads to the corollary that carbon fibre —or, as we know we should now say, carbon fibre reinforced plastic (CFRP)— should be used only where it fulfils a function so much better that the price difference is justified. It is not enough to identify where carbon fibre is useful; we must identify where it is unique. Yet when we compare its basic engineering properties with those of other materials we find in fact nothing unique about any one property in isolation. Contrary to popular opinion, CFRP has a lower ultimate tensile strength than has steel, while its Young's Modulus is, for instance, less than half that of tungsten carbide. The wear and frictional properties of CFRP are striking and have been the subject of detailed investigations,[2,3] but have not so far in themselves proved to be a sufficient justification for any identified application of CFRP. For it is in properties in combination, and not in isolation, that the uniqueness of CFRP is to be found. It is a philosophical maxim that the uniqueness of the individual lies in that combination of characteristics which is peculiar to him alone; this is as true for an engineering material as it is for the individual person. 'Job finding' in either context offers obvious analogies, and the various combinations of 'talent' leading to the selection of CFRP from among other candidate materials are most conveniently summarized in the algorithm of Fig. 5.1.

In this figure the most direct route to CFRP is 1N-2Y-3Y-4N (i.e., the answers to the first four questions are, in order, 'no', 'yes', 'yes', and 'no'), and certainly the uniquely high stiffness/weight of CFRP has led to more uses (table 5.2) than any other single property combination. However, to make the point that other 'bonus' property combinations are, in aggregate, just as powerful in selecting CFRP; such applications are listed in a separate table, 5.1. Here uses exploiting both stiffness and low density are explicitly omitted, although usually one or other of these properties is critical. Further, it has been acknowledged that more than two properties are frequently needed to constitute a unique combination; in particular, the strength/weight ratio of CFRP is assumed to be so little superior to GRP that at least one further property must be adduced. We say 'assumed' because there are good theoretical grounds for supposing that the strength of carbon fibre will be increased in the foreseeable future, and because in applications like containment vessels

161

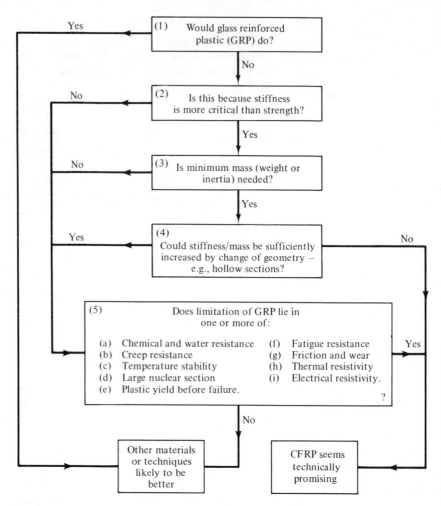

Fig. 5.1 *Property requirements leading to the selection of CFRP.*

the effective strength of glass can be limited by the breaking strain of the matrix.

It is remarkable that the applications listed in table 5.1 are not only nearly as numerous as those in table 5.2, but on balance may well account for the larger usage of carbon fibre over the next decade. Nonetheless, stiffness/weight is the most powerful single property combination, and is, after all, the *raison d'être* of carbon fibre: for it was the predictions of what specific stiffness should be available from carbon that inspired the RAE workers of the 'sixties to close the gap between the physicist's theory and the engineer's reality. An earlier review[4] analysed potential applications according to the constituent

Table 5.1 *Applications of CFRP exploiting properties other than specific stiffness*

	Stiffness	Low Density	Strength	Friction and Wear	Creep Resistance	Chemical/Biological Inertness	Fatigue Resistance	Absence of Plastic Yield	Temperature Stability	Small Nuclear Section	Non-Magnetic	'Logic Path' in Fig. 5.1	Reference
Drum Slats	×					×						3N–5(a)	59
Gas Centrifuge		×	×			×						2N–5(a)	43
Denture Palate			×			×						2N–5(a)	44
Wire-Bunching Bow		×	×	×	×		×		×		×	2N–5(b, f, g)	47
Flywheel		×	×		×				×			2N–5(b)	52
Irradiation Capsule			×							×		2N–5(d)	65
X-Ray Dosimeter			×							×		2N–5(d)	54
Prostheses		×	×					×				2N–5(e)	24
Sports Equipment		×	×	×	×			×	×			2N–5(b, c, e)	33
Sports Car Body	×							×	×			2Y–3Y–4Y–5(f)	50
Compressor and Pump Vanes	×			×				×	×			2Y–3N–5(c), (e)	48
Measuring Calipers	×				×				×			2Y–3N–5(b, c)	51
Radar Dish	×							×	×			2Y–3N–5(c)	50

parameters of specific stiffness, namely: mass, volume, deflection and the force causing deflection. This analysis is repeated in table 5.2, along with the other classifications given in the column headings. These will be explained in some detail before describing the applications in the line headings; indeed only an outline description will often be necessary once the nature of the entries in table 5.2 is accepted.

Reason for use. We are asking why we must answer 'no' to question 4 of the algorithm of Fig. 5.1, that is why, in flexural stiffening it is not possible to redistribute the existing material further from the neutral axis. Answers are coded thus:

Table 5.2 *Applications of CFRP exploiting specific stiffness*

	Reason for Use	Merit Index (See table 5.3) Primary	Secondary	Bonus Properties	Deflection — Buckling Failure: Strut	Panel	Fluid Dynamics	Geometry	Force: Constant	Variable	Mass: Weight	Inertia	Cost: Economic	Sporting	Military	Aesthetic	Reference
Analytical Centrifuge	1	M1	M5	b				×	×		×	×					
Generator End-Bell	1	M1		b, h				×	×		×	×					49
Transmission Shaft	1	M1	M5	f				×	×		×	×	×	×			
Springs	3	M1	M6	b, c, e, f				×		×	×	×		×			7
Compressor Blades	2	M1	M6	a, b, c, e, f			×	×	×		×	×					18
Antennae	4	M4		b, c, i				×	×		×	×	×		×		
Fishing Rod	2, 4	M6	M1		×		×				×	×	×		×		46
Levers	3	M1	M5	c, f, g				×	×		×	×					
Heddle Frame	2	M1	M5	e, g				×	×		×	×					8
Train Pantograph	2	M1	M5	a, f				×	×		×	×					53
Push Rod	2, 3	M2		c	×				×		×	×	×				50
Telescopic Crane	4	M4		b	×			×	×			×					
Bicycle Frame	4	M2	M4	c	×			×			×	×		×			8
Spinnaker Boom	4	M2		a	×				×		×	×		×			57
Sailing Mast	2, 4	M1	M2	a	×			×	×		×			×			57
Dinghy Centreplate	2	M3		a, e			×	×	×		×			×			
Loudspeaker Cone	1	M6		b						×	×	×				×	4

164

1   The stress is tensile and its direction coincides with that of the critical deflection. The first consideration means that there is no question of buckling failure, and the second that the deflection equates to the elongation, and so depends on the total area of the cross-section but not its shape.

2   In the existing designs the material is already situated as close as possible to the boundary defined by an outside constraint. For instance in fluid dynamics—aerofoil or hydrofoil sections—the boundary geometry is clearly fundamental to function, which may have to be compromised if thickness is increased to augment stiffness.

3   The complication of making hollow sections (which would be either left empty or filled with a light material to take any secondary stresses such as shear), is not worthwhile either because other properties of CFRP give support for its use anyway, or because the component is too small for raw materials cost to be of prime importance.

4   Hollow sections are already being used, but while there is no geometric limitation on the boundary dimensions, the corresponding further decrease in wall thickness would lead to local failure either from small transverse forces or from buckling under the primary forces.

Merit index. This is a quantitative measure of the performance of a material under given conditions. For the applications in table 5.2, the index can be more precisely defined as the ultimate load that can be withstood by a given structural form of given mass. As will be seen from table 5.3, in many cases the index equates simply to the specific stiffness E/ρ. But often a more complicated expression applies, notably where, as described in 4 above, there is no geometric constraint on the section boundaries. The values in table 5.4 are calculated relative to CFRP $= 1$, and perhaps predictably other materials are shown as being up to ten times less effective. But it has been truly said[5] that these indices are good servants but bad masters, and the following points should be borne in mind when studying table 5.3 :

(*a*)   The index necessarily ignores all but the most critical property relationship. In practice there may be different indices describing other relationships of secondary but almost equal importance: it may also be that the critical properties are non-quantitative, such as chemical resistance and machinability.

(*b*)   The usual caution about the anistropy of CFRP should be stated here. The indices quoted for the reinforced plastics (including glass and boron, as well as carbon) refer only to properties measured along the fibre direction in a unidirectional composite, assumed for the purpose of comparison to contain equal proportions by volume of fibre and matrix. By contrast the indices for the metallic materials are valid for virtually any direction of measurement. However, this caution can be over-emphasized; study of tables 5.1 and 5.2 witnesses to the predominantly unidirectional nature of the stresses

Table 5.3 Merit indices

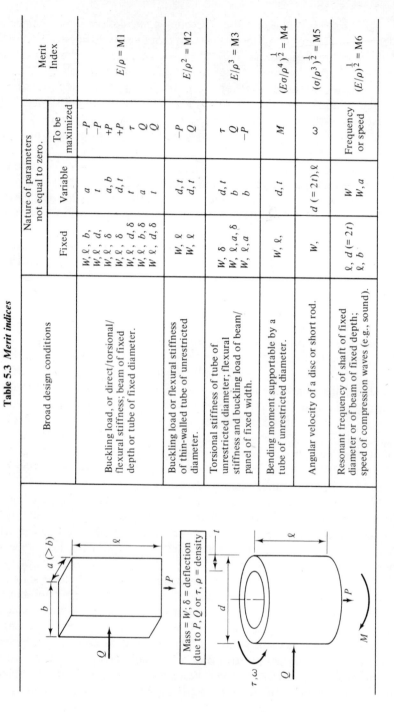

Mass = W; δ = deflection due to P, Q or τ, ρ = density

Broad design conditions	Nature of parameters not equal to zero. Fixed	Variable	To be maximized	Merit Index
Buckling load, or direct/torsional/flexural stiffness; beam or tube of fixed depth or tube of fixed diameter.	$W, \ell, b,$ $W, \ell, d,$ W, ℓ, δ W, ℓ, δ W, ℓ, d, δ W, ℓ, b, δ W, ℓ, d, δ	a t a, b d, t t a t	$-P$ $-P$ $+P$ $+P$ τ Q Q	$E/\rho = \text{M1}$
Buckling load or flexural stiffness of thin-walled tube of unrestricted diameter.	W, ℓ W, ℓ	d, t d, t	$-P$ Q	$E/\rho^2 = \text{M2}$
Torsional stiffness of tube of unrestricted diameter; flexural stiffness and buckling load of beam/panel of fixed width.	W, δ W, ℓ, a, δ W, ℓ, a	d, t b b	τ Q $-P$	$E/\rho^3 = \text{M3}$
Bending moment supportable by a tube of unrestricted diameter.	$W, \ell,$	d, t	M	$(E\sigma/\rho^4)^{\frac{1}{2}} = \text{M4}$
Angular velocity of a disc or short rod.	$W,$	$d\,(=2t), \ell$	ω	$(\sigma/\rho^3)^{\frac{1}{2}} = \text{M5}$
Resonant frequency of shaft of fixed diameter or of beam of fixed depth; speed of compression waves (e.g., sound).	$\ell, d\,(=2t)$ ℓ, b	W W, a	Frequency or speed	$(E/\rho)^{\frac{1}{2}} = \text{M6}$

Table 5.4 *Merit indices relative to type 1 CFRP*

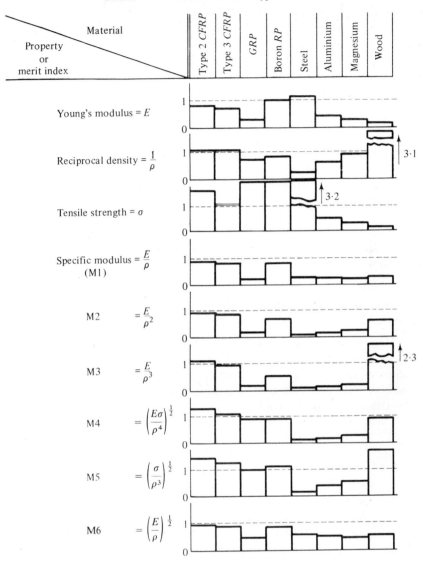

in components to which CFRP lends itself anyway. We shall see that these are either critical engineering artefacts designed to withstand one well-defined source of stress, or the more sophisticated types of sports equipment—in either case they are designed for use rather than abuse.

(c) Some qualification must be made on the dramatic relative index quoted for CFRP in circumstances which allow the section boundaries to be

167

extended from the neutral axis until it is local buckling of the wall that becomes critical (case 4 above). For, as an alternative to choosing a stiffer material, the sectional material can in turn be re-distributed about the local axis of buckling. Consider for example the simple tubular strut, where there is no limit on the diameters, the compressive force is given, and the weight is to be minimized. An outer diameter can be calculated at which buckling of the whole tube and at some random point in the wall would occur simultaneously; the weight corresponding to these diameters would be the minimum for a simple tube, but would be greater than that necessary using concentric tubes joined with a shear-carrying plastic foam to give a sandwich-wall effect. As an academic exercise this technique for guarding against thin-wall failure can be extended indefinitely: in practice one clearly reaches an early limit where the material cost of CFRP is overtaken by the production cost in the more complicated geometry.

Subject to the above reservations, the merit index remains an important concept in optimizing material usage. We may expect to hear more of it as materials science feels the impact of the computer, and of Lord Kelvin's adage that quantitative measurement is the basis of understanding.[6]

Bonus properties. These are coded by the letters used in question 5 of Fig. 1.

Deflection. This is analysed according to its criticality in elastic failure or to some geometric consideration.

Force. Only one force is assumed to be critical for design purposes, and may be either constant or variable. In practice this means that in all structures, including structural members of mechanisms, the design calculation is one of statics; only where the component is a mechanism in itself (e.g., a spring), need the variability of force for design purposes be observed.

Mass. This may be critical because of gravity (i.e., weight): in a moving component it is inertia, either in addition or as an alternative to weight, that is critical.

Cost. Making a component in CFRP may be justified by simple economics. In particular, the capital cost of a machine is meaningful only when compared to its productivity—if we denied this, we should logically mow our lawns with nail-scissors. Alternatively, in any competitive situation Man puts an enormous premium on anything which gives him the edge, however slight, over his competitors. On the battlefield, the modern infantry platoon with the lighter equipment is, other things being equal, the one that will win. Analogous arguments apply to sportsmen and to perfectionists in the fine arts.

The order in which applications are listed in tables 5.1 and 5.2 has been determined by continuity of property exploitation, and not by area of application; this has resulted in such diverse uses as the gas centrifuge and dentures appearing as bedfellows. This treatment is deliberate, in order to avoid the syllogisms that arise, for instance, through supposing that because CFRP can gainfully replace steel in certain shafts it can necessarily also be used successfully for the gear wheels with which the shaft is fitted. The design criteria for gears are quite different from those for shafts, and the fact that CFRP has in reality been tried for both is coincidental rather than self-evident. Until the engineer recognizes the error of such thinking, he will always be subject to unnecessary and wasteful disillusionment with the new materials available to him.

However, the engineer normally specializes by industry, not by the materials he uses, and therefore in the descriptions which follow the order of tables 5.1 and 5.2 has been abandoned in favour of classification by area of application. Such classification is necessarily arbitrary and at first sight indicates that the applications discussed occasionally trespass outside the strict definition of 'mechanical engineering'. The 'production' and 'process' industries clearly have a large area of overlap, and the various compressor applications discussed under the latter heading could arguably have been considered elsewhere. The justification is that in the process industries these applications have the greatest attraction, the need for chemical resistance being superimposed on the premium deduced for CFRP from purely mechanical requirements.

5.1 Production Engineering

It has already been observed that the capital cost of a machine assumes meaning only when compared to its productivity. Even this is an over-simplification, ignoring such important factors as utilization and interest rates. However, the capital cost will be little altered for any machine in which carbon fibre is likely to be introduced, for the simple reason that so little of the new material will be needed. For in the great bulk of machinery, weight is rarely a disadvantage, and stiffness is less critical than strength, especially impact strength and resistance to crack propagation. The most promising of the limited areas where CFRP can be exploited is in fast-moving components whose speed is limited by their own inertia rather than by extraneous forces. Production machines handling light media therefore offer frequent opportunities for the cost-effective application of CFRP. Examples are packaging, printing, gas circulation, and the handling of textiles and light wires. Two specific instances are given in this section, but a number of uses described under later headings could be classified within this generalization by regarding as 'light media' electricity, vacuum, and even 'information'—this latter to cover applications in data processing.

5.1.1 THE HEDDLE FRAME

The first and probably most published industrial application of CFRP was the reciprocating component which alternately lifts and lowers the warp in

169

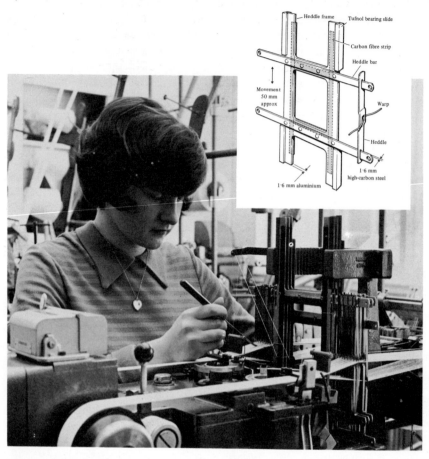

Fig. 5.2 *CFRP heddle frame (courtesy Bonas Brothers Ltd).*

certain narrow-weave looms (Fig. 5.2). Because, as shown, several frames are used in series, there is a geometric limit on the stiffening that can be obtained through thickening the section of the frame, which in any case is rather too small and delicate to make this solution worthwhile. The inertia and stiffness of this relatively small component dictate the productivity of the whole machine, and the cost/effective argument is irrefutable—a potential doubling of speed in return for a very few pounds worth of carbon fibre. In fact the new frames cost some ten times more than the originals, but this represents an increase of only $2\frac{1}{2}$ per cent on the price of the whole loom. An intermediate productivity increase of 50 per cent was obtained by replacing only part of the aluminium of the original frame with CFRP, illustrative of a general principle.

170

For undeniably with CFRP there are problems of designing joints, holes and indeed anything but the most simple free-standing shape. But a frequently satisfactory compromise is to insert CFRP strips into the simply stressed regions, retaining the original materials in the regions of more complicated geometry and stress.[42] When, as in other instances, the CFRP is hot-moulded *in situ*, allowance has to be made for the slightly negative thermal expansion of carbon fibre, and in the aircraft world,[7] techniques are being developed not merely to meet this effect, but, by pretensioning, to exploit it. However, in general engineering components not subject to a large range of service temperatures, it is sufficient simply to stick on strips of pre-cured CFRP with a cold-setting resin, retaining some metal in parallel to provide a large shear-carrying area between the two materials. If the component is critical for strength as well as stiffness, it may be desirable to match the Young's Moduli by adjusting the fibre proportion in the CFRP. In practice a 60 per cent V/V type 1 composite has about the same stiffness as steel, so in a bar compounded of two identical strips of CFRP and steel bonded to each other, the stiffness is equal to an all-steel bar, but the weight is reduced by a factor of about $\frac{5}{8}$ (CFRP having roughly one-quarter the density of steel).

Of the many emerging applications elsewhere in the textile industry, actuating rockers,[8] needle holders,[8] and rapiers[9] have already been publicized.

5.1.2 WIRE BUNCHING BOWS

The manufacture of flexible conductors for electric cables requires the twisting of a number of individual fine wires into a bunched strand, with a regular pitch of twist. In the typical machine of Fig. 5.3, the configuration is such that

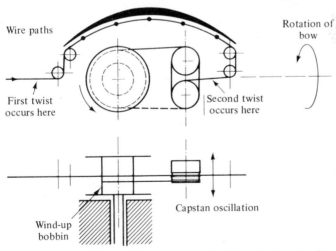

Fig. 5.3 *Double twist bunching machine (courtesy Trafalgar Engineering Company Ltd).*

one revolution of the bows imparts two twists in the wire and thus the production rate of the machine is directly proportional to the bow speed. The prime factor in the selection of a material for the bow is governed by the fact that stress in the component is determined by centrifugal force and thus the main property required is a high strength/weight ratio. It seems at first sight surprising that the choice was not glass reinforcement, with its greater capacity for storing strain energy. However, a comparatively low-modulus carbon fibre was found to offer adequate shock resistance as well as a number of critical properties not to be expected from glass. Amongst these are excellent mutual tolerance of both wire and bow when in abrasive running contact; stability of shape under varying temperatures and with the passage of time, following from the negligible expansion coefficient of carbon fibres and their negligible creep. The cost of the new bow is perhaps an order of magnitude greater than that of one in GRP, but since the bow is only one component, albeit a critical one, the cost of the whole machine is increased by a small percentage. However, in a typical machine the speed of rotation has been increased from 1000 rev min^{-1} up to a potential of 3000 rev min^{-1}. So the cable maker is offered a real bargain; a high increase in productivity in return for a small percentage cost increase.

Mention has also been made in the literature[59] of CFRP spooler fliers and flyer arms in machines for stranding fine wires.

5.2 Process Engineering

There has been an impressive growth in the use of GRP in the chemical industries, where massive glass and carbon/graphite have both long been accepted as standard materials. But before inferring that a general case for CFRP follows automatically, we must remember that the chemical resistance of a two-phase material is necessarily worse than that of its individual constituents. Thus CFRP does not offer the same across-the-board resistance as does either carbon/graphite or plastic alone. But additional constraints can dictate the selection of CFRP, as when a container for a glass-corroding fluid is subject to a variable stress. Under these circumstances, it is not sufficient to put a fibre-free lining coat on an RP structure, for the lining will rapidly develop microcracks which will progressively lead to complete failure. The answer is to reinforce the lining coat with a fibre that is both stiff and chemically inert. Pritchard and Henson[11] have shown that carbon fibre behaves admirably, and a combination of carbon fibre and brown asbestos has been reported[59] to have successfully displaced blue asbestos as the reinforcement of phenolic drum slats in certain special-purpose plant.

In the other areas of application described below the chemical resistance simply enhances the desirability of CFRP in what are essentially mechanical devices.

172

This is the machine through which a cost reduction factor of at least one-seventh has been forecast in the enrichment of uranium as a nuclear fuel. In conventional centrifuges the phases are usually liquid or solid, and so of a density comparable to that of the metal in which the separator bowl is normally constructed. Only when the phases are gaseous does the inertia of the bowl or cylinder become of critical importance in determining the stresses incurred and the consequent productivity of the machine. Since a fluoride is involved—in this case uranium hexafluoride—both metals and GRP require a protective coating which adds to the mass but not the strength, so for the optimum design CFRP is a 'must'. If this application survives the political scrutiny to which it is quite properly being submitted, then it has been estimated that by 1980 four million centrifuges a year will be needed to meet the European demand alone.

The merit index given in table 5.3 for a surface rotating in its own plane suggests that even in the absence of the chemical constraint CFRP would have considerable advantages for the cylinder ends. Experiments so far have yielded disappointing results, because of the difficulty in accurately orientating the fibre in the required uniform 'sweep's brush' fashion, and also possibly because of the more complicated stressing during changes of speed. However, once realized in practice, this advantage would extend to centrifuges handling conventional liquids, especially where the axial length of the separator bowl is small compared to its diameter.

5.2.2 COMPRESSOR AND PUMP VANES

The sliding vane for the eccentric rotary gas compressor (Fig. 5.4) was one of the first industrial uses of CFRP to be publicly mooted,[12] and a subsequent

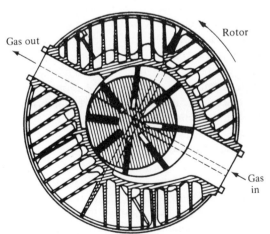

Fig. 5.4 *Principle of the vane compressor.*

173

design study[21] laid considerable emphasis on the theoretical savings in running costs that should accrue from the outstanding specific stiffness. However, the reaction from important sections of the industry has been that while stiffness is certainly critical, weight is rarely so. Preference for CFRP here rests more on either self-lubricating characteristics (offering a fail-safe provision for dry-running even where oil lubrication can be tolerated), the long fatigue life, and on chemical resistance when corrosive gases are to be handled. Important, too, is the amenability of the thermal expansion coefficients to be 'tailored', if necessary in two directions independently. For, within the limits imposed by the required stiffness, the proportions of fibre can be adjusted so that the vane expands longitudinally at the same rate as the cylinder, while expanding in the thickness direction at the possibly different rate of the rotor. Thus optimum clearances should obtain at all temperatures, giving not only maximum volumetric efficiency but also minimum wear. Examination of Fig. 5.4 will show that the vane/slot clearance is especially crucial, any initial inaccuracy leading to failure either through jamming or the progressive interaction of frictional heat and wear.

5.2.3 COMPRESSOR BLADES

Here the chief attraction is the simple one of providing the necessary cantilever and torsional rigidity for least inertia. Under this heading we can include blowers for gas circulation, where in process engineering a corrosive environment may corroborate the case for CFRP, as well as the first stages of the compressor on a gas turbine. Of the airborne variety it has been observed[13] that 'the advantages of carbon fibre are so enormous, 510 kg weight saved per aircraft, increased efficiency and specific fuel consumption, half the production costs of solid titanium blading, reduced containment problems, that CFRP fan-bladed aircraft are bound to become a reality'. Although aircraft engineering is outside the scope of this chapter, this quotation is relevant here because all the advantages claimed apply at least partially to the water- or vehicle-borne turbine, while even the land-based version exploits three of them and must hold out the same promise for CFRP in the rather longer term. At least one European manufacturer foresees that improvements in the temperature stability of resins should extend the use of CFRP blades to steam turbines by the 'eighties.

5.3 Electro-mechanical Engineering

Under this broad title we will discuss uses of CFRP which, although within the electrical, electronics, or communications industries, are primarily mechanical in nature and do not, notably, utilize the electrical conductivity of carbon. We will also consider here CFRP springs and levers, since although these clearly are of wider application which will take us outside electro-

174

mechanical engineering, their initial use will be in control and data-processing mechanisms. These are fundamentally machines handling information rather than mechanical power and so the component speeds are normally inertia-limited.

5.3.1 SPRINGS

The usual basic function of a spring is to store energy, which in any design—leaf, helical, or coil—is proportional to the square of the strength and the reciprocal of the stiffness. Where mass is critical the index by which materials may be compared becomes $\sigma^2/E\rho$, and since the most outstanding single characteristic of carbon fibre is the high value of its 'E', CFRP springs in this context have nothing to offer. Even where allowable deflection as well as mass is critical, the merit index becomes the specific strength σ/ρ, which, as has been pointed out, is only marginally greater for CFRP than for many conventional materials.

However, Henney has pointed out[7] that a flexural spring consisting of a layer of CFRP sandwiched by approximately equal thicknesses of GRP has, for the same stiffness, a weight about 15 per cent less than an identical spring entirely in GRP, which in turn stands well above conventional spring materials. As illustrated in Fig. 5.5, this principle can be extended by building a compound sandwich exploiting the spectrum of strengths and stiffnesses exhibited by different grades of both glass and carbon. The configuration shown offers a weight saving of 20 per cent, which figure could be pushed a few per cent higher by exploiting the infinitesimally variable ultimate tensile strength and E that can in principle be obtained in carbon fibre. In any case,

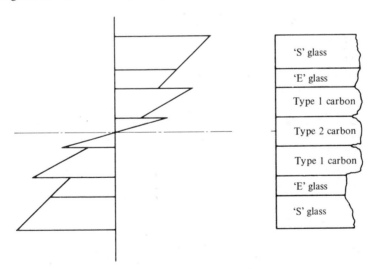

'S' glass
'E' glass
Type 1 carbon
Type 2 carbon
Type 1 carbon
'E' glass
'S' glass

Fig. 5.5 *Stress distribution in multilayer glass/carbon RP spring.*[7]

175

a small amount of carbon fibre, sited near the neutral axis where it will not be over-strained, can be expected to confer on a basically GRP spring an improved creep and fatigue performance, as well as increased temperature stability. In a word, such a spring should be more resilient, that is, it should yield its stored energy and return to its unstressed condition more reliably.[14]

All this said, maximum stored energy is by no means the universal criterion of spring design, or of components which have to exhibit some spring-like quality. Indeed, in the special case of the archery bow, it has been shown[15] that any energy in excess of that required to be imparted to the arrow is dissipated as kinetic energy in the bow limbs, and represents a waste of the archer's effort. In more general cases, the spring is required to exhibit a given force at a given deflection for the minimum of weight, so the merit index is simply specific stiffness. Snap fastenings on airborne electronics, access panels, and containers are obvious if apparently trivial examples. Potential applications are by no means limited to the simple leaf spring; for example, helical springs can be readily manufactured by filament-winding and offer a range of travel, rate, and lightness outside the design envelope of even the coiled tube construction in conventional materials.

However, it is in inertia-critical rather than weight-critical terms that the larger merits of a CFRP spring become apparent, since the frequency of natural vibration varies as the square root of the specific stiffness. One potential application exploiting this phenomenon is the ledger spring used as one element in a chain-bar mechanism in certain production machinery. It is in effect a flexible cantilever upon whose natural frequency the productivity of the whole machine depends. As in static bending cases, before selecting CFRP it is prudent to see if any saving of mass with existing materials can be achieved by alteration of the section geometry. However, in most cases the ledger spring is required to operate to its safe limit of elastic strain, and since this is no less for CFRP than for typical metals,[16] it follows once again that the availability of the new composite enlarges the envelope of characteristics available to the designer.

The 'root specific stiffness' rule is not confined to such simple macroscopic effects, but applies equally to any form of elastic vibration, be the wave motion longitudinal, transverse, or torsional. An extreme example is the ultrasonic wave collimator suggested by Tewary and Bullough,[17] who have pointed out that the angle through which such waves are refracted on passing from one medium to another depends on the relative specific stiffnesses of the two materials. Less esoteric applications of the same effects seem possible in machine tools, both to suppress noise and to reduce chatter, although the latter would only be realized by converting a substantial proportion of the structure to CFRP.[58] A tangible possibility is the boring bar for turning deep internal diameters. The frequency of longitudinal vibrations, such as sound, is a unique function of specific stiffness, and cannot be altered by change of geometry. Hence CFRP has been mooted as a propagator of sound in the loudspeaker diaphragm,[4] and actually used as a dissipator of sound in the jet engine[18]—an interesting example which also exploits the relatively good damping of CFRP and which has survived the turbine blade controversy without attracting much publicity. While admittedly there are many approaches to the problem of vibration control, it is suggested that this aspect of specific stiffness has been insufficiently recognized in the search for outlets for CFRP.

5.3.2 LEVERS

Many of the comments on springs apply equally here; indeed since the case for CFRP is strongest in high-speed components, resilience to repeated impulse loads is a prime requirement. Equally the uniquely high natural frequency of CFRP offers the designer wider scope of 'detuning' unwanted vibrations. One potential application is the pantograph mechanism which carries the overhead conductor on electric railways. Although there is no question of using carbon fibre to make the electrical connection, the mechanism as a whole transmits energy electrically rather than mechanically, so inertia forces predominate and must be minimized to obtain the most uniform contact pressure. The ultimate in stiffness is probably not needed, but by introducing a selective proportion of carbon fibre into a basically GRP construction random vibrations may be suppressed. In the aircraft world this principle has been well demonstrated in the helicopter blades that went into production in 1971.[19]

Another promising area is in data processing. Here it is commonplace to say that speeds are limited less by the electronics of the central processor, than by the mechanics of the peripherals. Promising areas of application of CFRP are, therefore, levers in print-outs and card sorters, as well as the recording head supports of random-access disc stores.[20]

5.3.3 INSTRUMENTATION

In principle CFRP has as much to offer here as in data processing, for the two areas overlap. At least one company is known to have investigated CFRP for a transducer component, but the overall response from the industry has been disappointing, probably because of the difficulty of making thin sections from carbon fibres currently produced.

The one published application in this category primarily exploits a quite different property of carbon fibre, namely its coefficient of thermal expansion. The 'tailorability' of this property of expansion has already been referred to; possibly more significant than the negative sense of the coefficient is its numerical smallness. By suitable choice of fibre/resin ratio, and of fibre direction, one obtains a plastic competitor to Invar. The measuring callipers shown in Fig. 5.6 are part of a dynamic system, so that low density is of some

Fig. 5.6 *CFRP measuring calliper (courtesy of National Engineering Laboratory).*

importance, but the crucial consideration is dimensional stability. This is the first known application so far to rely primarily on this unique property, but investigations are under way in a number of fields of metrology, including horology.

5.3.4 RADAR DISH

Three-dimensional curves are, broadly, easier to produce in plastic than in sheet metal, and stiffness is clearly critical to the accuracy with which the radar impulses can be beamed. Accuracy is also easier to achieve if the fibre has a high thermal conductivity, since even with a cold-curing resin lack of symmetry in dissipating the exothermic can cause distortion.[60] Thus even when mass is not particularly critical (as in civil ground-borne equipment which needs to scan at only moderate speed) there is some argument for CFRP, although generally a sufficiently thick section of GRP provides a cheaper solution. However, neither GRP nor any conventional material will give a distortion-free dish in, for instance, desert conditions where a temperature difference of over 50°C can readily develop between the sunny and shaded faces. Thus the virtually zero expansion coefficient of carbon fibre is sufficient justification of its use. Because the thermal expansion in directions other than that of the fibre length is large, doubts might reasonably be felt about the practical achievement of a distortion-free dish, but the Royal Radar Establishment has shown such doubts to be groundless. After detailed

178

consideration of fibre orientation several trial dishes have been produced and passed field trials with flying colours. Because, unlike glass, carbon is opaque to microwaves there is no possibility of a CFRP radome, mechanically attractive as it would be.

5.3.5 ANTENNAE

There is no general requirement for especially light materials in stationary lattice structures, since by comparison to self-weight the external forces are large and are the more responsible for the critical design stresses. But when, as in aerials, there is no functional weight to support, but only windage and self-weight, both slim sections and low density are at a premium. For high frequency transmissions, such as microwaves and radar, there is also a requirement for rigidity, so all factors added together present a very strong case for CFRP, provided its opaqueness to microwaves can be accommodated or even, for screening purposes, exploited. For often provision has to be made to screen off 'side loops' from a radar transmitter, as such loops can not only cause spurious signals, but also be a lethal hazard at close range.[21]

5.3.6 GENERATOR END-BELL

Applications in heavy electrical engineering will, as we have seen for heavy engineering generally, be relatively few. One promising use is a mixed glass/carbon fibre circuit breaker, where presumably the virtues described earlier for the glass/carbon spring are compounded with corrosion resistance. The one published use relates to the end-bell or retaining ring fitted on large electrical machines to restrain the end-windings against centrifugal force. It has been reported[49] that to make generators of output greater than 1350 MW the stiffness/weight ratio of conventional materials will be inadequate, and so CFRP will be a *sine qua non*. But it will be seen that the end-bell is also included in table 5.1, because at least one large European manufacturer sees a more general case for CFRP. He points to the problems with the materials at present used for end-bells: the structurally preferable steels are ferromagnetic, and GRP, despite its advantages (in this case) of being neither magnetic nor conducting, suffers from lack of stiffness, poor creep and a large coefficient of thermal expansion. On all three points CFRP compares favourably. Whereas low-E materials can often be made to give adequate stiffness in the bending mode through the appropriate section geometry (e.g., channel or tubular construction), no equivalent opportunity exists when, as in the present case, the critical stress is unidirectional. Another attraction is the selectivity of modulus that can be offered in carbon fibres, by varying the pyrolysis temperatures employed during manufacture. Ideally the thermal expansion, as well as the modulus of the retaining ring material should equal that of the materials retained, and though in this respect CFRP is less than ideal, it is better than GRP in both magnitude and direction of

179

expansion. More important, the creep data established so far show carbon to be far superior, and even though by definition exhaustive long-term performance has yet to be established, considerations of the fundamental stability of carbon in any form lead to reasoned optimism.

5.4 Medical Engineering

Where carbon fibre and medicine have been mentioned in the same sentence, it is usually with the suggestion of CFRP implants. But since here chemical and biological compatibility are the first consideration, single-phase materials are normally to be preferred. For while there is considerable experience with all-carbon and with thermoplastic implants, the same is not true of the thermosetting resins with which most work on carbon fibre has been carried out.

Exceptionally, acrylic resins have long been used for denture palates, but a significant proportion of the denture-wearing public succeeded in breaking them at intervals. Attempts at reinforcement with glass were abandoned because of mouth irritation. Carbon fibre has not only eliminated this but brought as a bonus a small but appreciable saving of weight in the upper palate.

Probably the biggest benefit carbon fibre will bring to medicine will be indirectly, through electronics. Another interesting possibility is the harnessing of the wave collimation effect, already described, as focusing devices in the ultrasonic techniques recently adapted in surgery.[22] If feasible, such devices would clearly have a usage wider than in medicine, as is true for some of the other applications which will, for the purposes of classification, be considered below under the medical heading.

5.4.1 ANALYTICAL CENTRIFUGE

In discussing the uranium enrichment device (p. 173), it was indicated that the specific properties of a centrifuge rotor are critical only when the media being separated are of low mass; even then it must be admitted, specific strength rather than stiffness is critical, since shape changes in the rotor can normally be accommodated in the stator design. An exception arises in mass chromatography, where samples are analysed according to minute differences in the densities of their constituents. Blood samples, for instance, are placed in test tubes and centrifugally fractionated in this way, giving an asymmetric loading which imposes a stiffness requirement on the need for low inertia in the rotor. In the gas centrifuge a compound aluminium/CFRP rotor is envisaged as a developmental stage, and a similar approach seems sensible for the blood analyser. Here at the moment the preferred material is titanium, which is the easiest metal to compound with CFRP, because of its closely similar stress/strain[7] and electrochemical[23] characteristics.

180

5.4.2 PROSTHESES

The most advanced work in this field has been on support frames for thalidomide children. Such frames need to be strong, stiff, yet light, while the high labour content of such custom-built items reduces material cost to secondary importance. But the decisive characteristic in relegating other materials in favour of CFRP is its absence of plastic yield. For one of the two research teams[24] feels that the consequences of brittle fracture—which are mitigated by retaining the outer thermoplastic sleeve in which the CFRP is moulded— would be less serious than those of a permanent set which would cause pain to the child but escape notice by the adult eye. Admittedly a second team disagrees with this,[25] and in this application above all others, no-one would wish to use a new material purely for the sake of a design exercise. While case experience is settling the issue, engineers may draw a technical moral: most literature on CFRP treats lack of plastic yield as an unmitigated disadvantage, although Pearce has shown[16] that this is not so great as it might appear at first sight. However, in the present case, as in many engineering applications, brittle failure is possibly to be favoured over deformation—any ship's captain, faced with the choice, would rather have his rudder broken off than jammed. 'Advantage' and 'disadvantage' are emotive words, and in this basic process of matching materials to applications, it is important that neither should automatically be equated to 'property'.

Artificial limbs represent another promising area for CFRP, but because toughness is almost certainly needed here, the best solution is probably to use carbon fibre in conjunction with either GRP or aluminium.

5.4.3 X-RAY DOSIMETER

The device in Fig. 5.7 has been designed by the National Physical Laboratory for calibrating instruments used in Health Physics for monitoring low and medium-energy levels of x-radiation. Sensitivity to low-energy waves requires that the wall materials should have low x-ray absorption, which is approximately proportional to the fourth power of the atomic number. Now carbon has the fourth lowest atomic number of all solid elements, so CFRP is uniquely suitable for the circular hoops which provide the thin frame for steadying the inflated rubber-walled ionization chamber. A similar but less esoteric application exploiting the same property combination is the radiotherapy couch. Body tumours are now being treated as well as detected by x-rays, but since the equipment needed for treatment is heavier and more expensive than that for detection, there is an economic case for moving the patient under the treatment generator only after detection and location have been completed in a separate unit. This requires a couch that is stiff, light, and transparent to x-rays, a property combination in which CFRP stands unique. It is interesting that while carbon is opaque to microwaves, glass is largely opaque to x-rays but transparent to microwaves.[21] Thus, in their

181

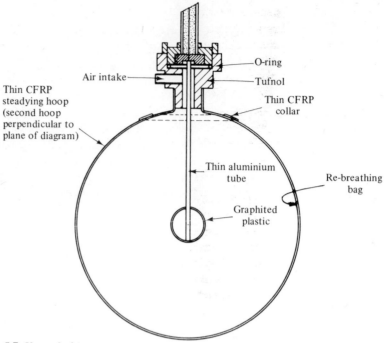

Fig. 5.7 *X-ray dosimeter.*

reaction to electromagnetic waves as often elsewhere, glass and carbon fibre are seen to be complementary rather than competitive.

5.4.4 IRRADIATION CAPSULE

One of the first non-aerospace uses of CFRP to be published was a capsule used to convey samples into a nuclear reactor for activation. Of course, this technique is of much wider application than in medicine and biology. To control the dosage the transfer in and out of the pile has to be as rapid as possible, and a pneumatic 'rabbit' system is used. Clearly strength/inertia is critical, and some sort of fibrous reinforcement is called for. Now while all reports to date[26] have shown little or no mechanical advantage of carbon over glass when, as here, the fibre orientation is random, carbon does offer the unique advantage of a negligible neutron cross-section. This means that the fibres neither sensibly diminish the neutron energy, nor themselves become activated and cause unwanted background radiation in subsequent analysis.

5.5 Automotive Engineering

Materials for cars, broadly speaking, have to be tough and cheap, so no wholesale use of CFRP can be foreseen. Although carbon fibres have already been successfully used in sports car bodies, it is in the engine and transmission that the ordinary motorist is likely to find benefit. He will probably not be conscious of the introduction of CFRP, any more than he was of the introduction of other specialist materials like silicone rubber or vanadium. CFRP piston rings, for example, have shown promise in laboratory conditions,[27] and feasibility studies have been made[28] of a CFRP connecting rod and piston. Because of desirability of a split construction in the first case, and high temperature in the second, the best approach would seem to be that of a basically metal structure with CFRP inserts, as described earlier. Push rods and rocker arms, being essentially control mechanisms, have been successfully developed; the advent of the OHV engine restricts their motoring interest to the amateur enthusiast 'souping up' his special, but other sorts of piston machines (like the compressor) present promising opportunities.

5.5.1 SPORTS CAR BODY

One of the earliest published non-aerospace applications of carbon fibre was the upgrading of a basically GRP car body with a 'tartan pattern' of the black tows. The body of one racing car was in this way reduced from 69·4 kg to 40·2 kg, and went on to win twice at Le Mans. Only 1·36 of the 40·2 kg were carbon, so the additional materials cost was not inordinate. Now it might be challenged that the increase of stiffness comes more from the change of technique than the change of material, so that glass rovings laid down in a similar way would produce virtually the same gain in specific stiffness. Tests on flat panels tend to substantiate this challenge but no evidence has come to light to indicate that the fatigue life and resilience would be increased in the same dramatic way as with carbon fibres. Not only did the Le Mans-winning car body need less repair, but in lasting two seasons it gave at least twice the life normally expected of the conventional GRP body.

5.5.2 TRANSMISSION SHAFT

CFRP shafts have already been designed for helicopter tail units, and it can be expected that they will slowly penetrate into at least the competitive sector of the automotive market. The generalized merit indices in table 5.2 must be treated with more than usual reserve, since although the principal stresses are torsional, the main requirement for stiffness is axial, and the fibre direction has to be compromised accordingly. Nonetheless, one authority[29] has calculated that a CFRP shaft should weigh only 0·36 that of the aluminium equivalent. On ground-borne vehicles, CFRP will probably be better exploited by increasing the stiffness for a given weight, thereby easing the design constraint on the number and disposition of bearings. In long vehicles, such bearings have to be provided to stop shaft whirl, and if a stiffer material means they can be dispensed with, savings in initial and maintenance costs result.

183

This would at first sight seem an unlikely application for what is essentially a weight-saving material, and indeed where the function of the flywheel is to store energy or to provide directional stability the merit index is a unique function of strength, so conventional materials are generally to be preferred.

However, increased awareness of noise and pollution on the roads has revived earlier ideas of replacing the internal combustion engine with a flywheel, and conventional filling stations with electrical service centres. Since the problem is then to carry an adequate minimum of kinetic energy for a given weight, the merit index becomes specific strength, and fibrous composites begin to look attractive. Such flywheels are being investigated in the United States by Rabenhorst, who has now proposed various 'brush' configurations such as that shown in Fig. 5.8. As each 'bristle' is an independent

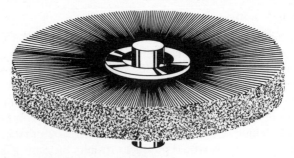

Fig. 5.8 *Circular brush superflywheel configuration.*

rod of composite weighing only a fraction of a gram, and as there is a small scatter of length around the nominal radius of gyration, any failure of the flywheel will be progressive, not abrupt. The design is thus inherently fail-safe, a characteristic which would probably become a legal necessity.[30] The design also side-steps the complex and possibly critical stresses that occur in a solid flywheel during velocity changes.

The argument thus far would show no advantage of carbon over glass reinforcement. An advocate for glass fibre might ironically concede that the property variability of carbon fibre for once could be turned to advantage in the fail-safe aspect, although he could expect to have his attention drawn to how much closer are the tolerances offered by carbon once in the composite state.[31] The more serious issues turn on how important is the greater energy stored elastically in the GRP wheel, and how acceptable would be its creep behaviour and limited fatigue life.

Mention of the racing car and motorcycle wheels is pertinent here both in area of application and some similarity of property exploitation. The wheels must be as light as possible, to reduce the unsprung weight and, in the motorcycle at least, the gyroscopic effect. The desirability of a safe-failure mode

184

supports the case for a fibrous composite, and as stiffness as well as strength is needed, CFRP is logically the preferred material. In so far as the stresses are produced by the centrifugal effect, this preference is supported by the entry in table 5.3 for a surface spinning in its own plane.

5.5.4 TELESCOPIC CRANE

These qualify for mention here as they are normally vehicle-borne. Here the geometry puts an exceptional premium on stiffness, and since the normal load is just a man with hand tools (e.g., for servicing large aircraft) and the crane has to 'train' in three dimensions, there is a strong case for CFRP, at least in the outer members. Similar arguments have been advanced for the uprights of a fireman's ladder.

5.6 Marine Engineering and Naval Architecture

In shipbuilding as elsewhere, CFRP will make its unassuming entry in detailed components rather than massive structures. Exceptionally the hulls of deep-submergence vessels represents an intriguing, but long-term, possibility, since the failure mode is buckling and weight-saving, perhaps surprisingly, is at a premium.

Many of the more immediate opportunities have been already covered, such as aerials and radar dishes, which form part of the topweight having an adverse effect on a ship's stability. All the arguments for automotive engine components apply equally here; indeed, if a CFRP con-rod could be shown to confer even the slightest economy of fuel, the fleet owner would lend a more attentive ear than the average motorist.

The Ministry of Defence (Navy) has carried out considerable research on carbon fibres and published a number of interesting papers, without disclosing any concrete applications. One of these papers[32] considers mixed glass/carbon composites, so it is natural to examine the case for using small quantities of carbon fibre in the GRP hulls of the new class of mine-hunters. Carbon fibres are non-magnetic, but so also is glass, so we must look elsewhere to identify the Navy's interest. This most likely turns on the resistance of CFRP to stress corrosion, already described under 'Process Engineering'. On the hull itself the effect of sea water in delaminating GRP is probably too small to be a consideration, but this is not the case with certain items of mine countermeasure equipment towed astern or over the side; since such equipment has to be man-handled, there is a case here for using carbon fibre for structural purposes as well.

The corrosion aspect has led to the suggestion of CFRP ship's propellers, but it is apposite and only fair to mention here the severe galvanic effect suffered by metals in the presence of carbon and an electrolyte. Even in the substantially dry environment of aircraft structures the limiting potential difference is 0·25–0·30 volts; this means that CFRP can acceptably be put in direct contact with titanium[23] but not with aluminium.[62] Although in CFRP the resin provides considerable insulation, the surface area of the fibres is large and special precautions are demanded, especially in a liquid environment where the carbon fibres are best exploited by putting them in the surface layer (p. 172). Because carbon is electro-negative, coming well below any of the metals commonly used in ships' construction (see Appendix), the established practice of providing sacrificial blocks of zinc or magnesium would be less effective.

In a ship's propeller CFRP would probably also suffer more than metals from cavitation effects, so almost certainly the optimal design would be a CFRP skeleton totally clad in bronze. Such a propeller would offer considerable weight-saving which, if of no particular merit in service, would offer the ship's operator some savings when airfreighting an emergency replacement, and a much bigger saving if the lighter unit enabled the replacement to be carried out by divers instead of in a dry-dock.

Uses of carbon fibre in small craft are included in the section immediately following.

5.7 Sports Equipment

This general title covers a multitude of ideas for CFRP, many of which, like fishing rods, depend primarily on stiffness/weight. Outside such applications one can state the requirements for elastic behaviour less certainly, and one is tempted to conclude, for instance, that there are as many specifications for the ideal golf club shaft as there are aspirant Open Champions! When asked to describe what mechanical properties he looks for in his equipment, the sportsman, especially the ball game player, will sooner or later use the word 'feel'; in so far as this word is capable of analysis, it probably means a variation of elastic behaviour, especially resilience, along the shaft or handle.[14,33] One reason for specifying carbon fibre here, then, is the amenability of its modulus to 'tailoring', by controlling the temperature of pyrolysis, but a stronger reason is insensitivity to change of temperature. GRP vaulting-poles, for instance, are the choice of champions, but elaborate precautions have to be taken to keep them at an even temperature between jumps. Given adequate breaking strain, in a CFRP pole the carbon fibre would both so dominate the overall composite behaviour, and also vary so little with temperature, that such precautions would be reduced to mere gamesmanship. Other sporting applications exploit other properties; friction and wear are obviously important in skis, whereas low density, electrical conductivity, and absence of permanent set would seem to make CFRP uniquely suitable for the modern fencing foil. Other applications which are already under assessment are described below.

186

5.7.1 BICYCLE FRAME

A successful frame lighter than any obtainable in conventional alloys has been manufactured. Track-racing cyclists are less interested in stiffness *per se* than in minimizing the strain energy capacity of their frames—a further instance of a usual drawback in a material being turned to advantage in certain cases. For the energy stored during the power peak of the cycle (in the abstract sense) is returned out-of-phase, and decreases the rider's control of his machine. In road racing simple weight reduction is the main attraction. Weight-saving is important in any form of competition, whether to give greater acceleration in sprints or less to manhandle in scrambles.

5.7.2 THE SPINNAKER POLE

This spar used in sailing represents a classical case of a pure strut. The diameter is chosen to meet the best compromise between the conflicting requirements of handleability and buckling strength, while the mass needs to be as small as possible to facilitate the all-critical manoeuvres of setting, lowering, and gybing the spinnaker sail. Mixed carbon/boron RP construction was used by the successful defender of the 1970 America's Cup for the spinnaker pole as well as the boom and cross-trees.

5.7.3 THE SAILING MAST

To take another example from the same sport, the design factors for the mast are different and more complicated. The lower part, below the shrouds, is primarily a strut, and although for ease of manufacture this might well be made in CFRP, in a keel boat weight is not really critical here. However, in the upper part of the mast the main demand is to control the curvature of the sail leach in a predictable manner, while offering the minimum of obstruction to the airflow around the crucial leading edge of the aerofoil constituted by a full sail. Mass is at some premium here, as it impairs the stability in light airs and when the sails are down. In a dinghy, this reasoning is reinforced by the need to minimize the all-up weight, which is more critical to speed than in the keel-boat.

5.7.4 THE DINGHY CENTREPLATE

This is one more possible application from the same sport. A combination of drag and water pressure on the leeward side conspire to twist the plate so as to impair its function of minimizing transverse motion due to the beam component of the wind. Minimum specified weights normally refer to the bare hull only, so the use of extra light-weight fittings like the mast and centreplate provide the only opportunity for the enthusiastic helmsman to steal a march on his opponents.

Oars for the British crew in the 1972 Olympics were made from conventional woods reinforced with longitudinal and helical stiffening strips in CFRP[64] and proved 10 per cent stiffer than the standard items yet 30 per cent lighter. The side face area was also reduced by 25 per cent, giving lower wind resistance and improved handling in rough water.

This is an outstanding example of a 'hybrid' composite, and it is valuable to speculate why the manufacturers, the GKN Group Technology Centre, should have chosen for their first experience to use CFRP in conjunction with wood, rather than with GRP, about which more has been written. While it is impossible to impute to the oar any one simple Merit Index, study of table 5.3 indicates that it is likely to have as a denominator a power of E greater than unity, i.e., a decrease in density is more valuable than a proportional increase in modulus. Table 5.4 shows that in such cases wood lies much closer to CFRP than does GRP, so that the virtue of CFRP is less compromised by an 'admixture' of wood than of the same proportion of GRP.

This completes our review of carbon-reinforced thermosetting resins. To conclude we will return to the premises set out at the beginning of the chapter, that carbon fibres are expensive, and can be used only for reinforcing thermosetting resins. We will continue to assume that thermoplastics are the only alternative matrices, ceramic, carbon, and metal reinforcement being left to other chapters. The challenges to the other assumptions turn out to be interrelated, and are best taken together against a background of the following facts:

(a) The most outstanding property of carbon fibre is its large specific stiffness, but this is only apparent in composites where the fibres are highly orientated and where in consequence the structural properties are highly anisotropic.

(b) The manufacturing techniques for thermosetting resins—hand lay-up, prepreg moulding, and filament-winding—lend themselves readily to accurate fibre collimation, whereas the various casting techniques used with thermoplastics do not.

(c) In any randomly orientated fibrous composite the strength and stiffness are only a fraction of those obtained in a unidirectional composite, and even on a weight-for-weight basis compare unfavourably with anisotropic materials like the conventional structural metals. With carbon fibre/thermoplastic compounds this disadvantage is aggravated by the lack of adhesion of the two phases, and the reduction of fibre length on moulding, two phenomena which do not affect the stiffness but do conspire in extreme cases to give a strength actually less than that of the unfilled thermoplastic.[26]

(d) Plastics compete with metals where accurate, complicated shapes are needed; the savings in the relatively expensive machining and forming costs

188

of metals can then offset the relatively greater raw materials cost of plastics. This is more frequently true for long runs in thermoplastics, or for short runs in thermosets, a notable example being the wire bunching bows described on p. 171. Despite the considerable weight of carbon fibre in each bow, the unit finished cost is reported[61] to be no more than that of the metal bow it replaced.

(e) In the discussion so far we have assumed the term 'carbon fibre' to cover only the high modulus varieties, the cost of which is largely dictated by the high temperature furnaces necessary to obtain the required degree of graphitization. Textiles which are simply chemically pure carbon can be made more cheaply, and are far less selective in choice of raw material, as many a housewife with an unregulated iron will testify. Apart from the stiffness and strength, such fibre can be expected to exhibit most of the other 'bonus' properties listed in the column headings of table 5.1.

Taking, as it were, the algebraic sum of the advantages and disadvantages, it appears that randomly orientated carbon fibre thermoplastic composites have nothing to offer in meeting purely structural requirements, but may be uniquely suitable in meeting other applications requiring one or more of the bonus properties; further, such properties may be obtained from the cheaper varieties of carbon fibre. A well documented[34,35] example is the medium-loaded gear wheel, needing to exhibit such inter-related characteristics as stiffness, thermal conductivity, dimensional stability, and low wear. Various thermoplastic/carbon fibre mixtures are showing promise. Another reported application of random carbon fibre, this time in a thermosetting resin, is for moulds and press-tools used in various metal-forming processes;[36] here thermal conductivity and abrasion resistance are the properties chiefly exploited.

But it must be conceded that no randomly reinforced carbon fibre composite components are known yet to be in actual production. This is primarily because of the number of variables that have to be examined once we break away from the purely structural applications. The gear-wheel, for instance, has taken three years development, because it is necessary not only to optimize such factors as matrix type, mould gate sizes, fibre fill, and shrinkage allowances, but to consider different approaches such as the use of an accepted self-lubricant compounded with a conventional structural material. And before substituting low modulus carbon fibres it would have to be confirmed that they do indeed offer the bonus properties required. Of the thermal conductivity and chemical resistance, for instance, there is no doubt, but equivalent bonus properties cannot be taken for granted. Giltrow[37] has shown that tribological conditions demanding type 1 are exploiting the actual stiffness of the fibre. So while low-modulus material would not serve here, it might well where type 2 is the preferred material, and this broadly is in a thermoplastic working against hard smooth surfaces and stainless steels.

189

There thus emerges the possibility of a corrosion-free, self lubricating, gear train with pinions alternating in stainless steel and carbon fibre reinforced thermoplastic (CFRTP), capable of transmitting considerable power.

If the development of such applications is tortuous, then the rewards will ultimately be significant, because, to repeat, the economies of thermoplastics are realized only in long production runs. The production cost of the CFRTP gear-wheel was long since[35] shown to be less than that of the steel equivalent, the apparently high cost of carbon fibre notwithstanding. So, by definition, a viable application of CFRTP will be a mass-production one, and this will in turn reduce the price, whether the low or high modulus grade be preferred.

For low-modulus fibre too, despite its longer history, is still poised near the top of the price/volume curve, although the publicity given to the high-modulus variety has, perhaps ironically, stimulated its growth.

In the aerospace world low modulus carbon fibre mixed with phenolic resins is the leading material for ablative rocket nozzles and nose-cones, while a similar material kilned to give a carbon/carbon fibre composite is a promising contender for aircraft brake units. Here the main attraction is the combination of high thermal conductivity, capacity and stability with a toughness greater than is offered by conventional carbon/graphite.

Outside aerospace, the main roles have been non-reinforcing ones, exploiting tribological properties, chemical and temperature stability, together with thermal and electrical conductivities not offered by any other textile. The biggest single use is for braided packings, mainly as pump gland seals where the speed, temperature, and/or chemical environment make conventional cotton or asbestos packings inadequate. Other uses are the insulation of vacuum furnaces, exploiting chemical and thermal stability, and as a non-corrosive conductor in the reverse electro osmosis technique[38] of damp-proofing buildings. These applications belie the original premise that carbon fibres are useful only for reinforcement, and if it is retorted that this premise was made only in the context of high modulus fibre, then at least two non-reinforcing applications of such fibre may be cited. These are the carbon fibre brush, described in chapter 6, and the oscillating fibre microbalance.[39] Here the weight of small bodies like pollen particles is determined by mounting them on the end of an individual carbon fibre and observing the change in resonant frequency, both specific stiffness and electrical conductivity being exploited.

Esoteric as it is, this application represents an extreme example of how usage of materials can be virtually independent of their cost, and we can allow ourselves the generalization that the growth of carbon fibre usage to date has been governed less by its cost than by knowledge of where and how to use it. Despite continuing confidence that modifications of present carbon fibre manufacturing technology,[40] or radically new methods[41] will continue to bring down the price, this will always probably be, on a weight-for-weight

190

basis, an order of magnitude greater than for glass. To this extent, then, our original premise must stand, but we are arguing that it is not of essence to the broad future of carbon fibre. The manufacturer must continue to optimize a complicated function of such inter-related variables as price, volume, and fibre property, especially modulus, but without the consumer's co-operation he has more unknowns than equations to solve them. The consumer must also recognize that he must usually himself do some development to bridge the gap between a good carbon fibre and a good design in CFRP. Rejection of this challenge will only strengthen the case for his competitors to accept it.

References

1 TIDBURY, G. H., and TETLOW, R., 'Carbon fibre for vehicle structures: Cranfield Institute of Technology Data Sheet 92', *Automotive Design Engineering*, **9**, 62-65, 1970.
2 PERKINS, C., *Tribological characteristics of carbon fibre filled thermoplastics*, Paper 12, Conference on Reinforced Thermoplastics, Plastics Institute, London, 1970.
3 GILTROW, J. P., and LANCASTER, J. K., *Properties of carbon fibre reinforced polymers relevant to applications in tribology*, Paper 31, International Conference on Carbon Fibres, their Composites and Applications, Plastics Institute, London, 1971.
4 BEDWELL, M., 'Understanding carbon fibre', *Materials towards the '70's*, 6–12, Engineering, Chemical and Marine Press, London, 1969.
5 HOWARD, H. B., *Merit indices for structural materials*, AGARD Report 105, Advisory Group for Aerospace Research and Development, Paris, 1957.
6 GORDON, J. E., *The new science of strong materials*, p. 22, Penguin, 1968.
7 HENNEY, A. S., 'Preliminary design of structural components in carbon fibre reinforced plastics and metals', *Aircraft Engineering*, **42**, 11, 18–24, 1970.
8 HEINS, G., *Practical applications of carbon fibre composites in general engineering*, Paper 5, Conference on Practical Applications of Composite Materials, PERA, Melton Mowbray, 1971.
9 Advisory and Components Services Ltd, *Reinforced Plastics*, **15**, 8, 190, 1971.
10 MARTIN, G., Trafalgar Engineering Ltd, Private communication.
11 PRITCHARD, G., and HENSON, J., 'Carbon fibre paper—an anti-corrosion material', *Composites*, **2**, 1, 6, 1971.
12 THOMPSON, R., 'Sliding vane type compressors and motors using carbon fibre reinforced plastics', *Design Engineering*, February 1969.
13 MORTIMER, J., 'Making the most of carbon fibre', *The Engineer*, **232**, 6016, 26–29, 65, 1971.
14 PHILLIPS, L. N., Royal Aircraft Establishment, Private communication.
15 WHITE, A. E. S., Morganite Research and Development Ltd, Private communication.
16 PEARCE, D. G., 'Designing for carbon fibre reinforced plastics', *Design Engineering*, February 1969.
17 TEWARY, V. K., and BULLOUGH, R., 'Carbon fibre composite as collimator and filter for stress waves', *Journal of Physics* D, **4**, 2, L5–L6, 1971.
18 MORTIMER, J., 'Demand for carbon fibre will continue despite the setbacks', *The Engineer*, **232**, 6015, 7, 1971.
19 PORTNOFT, A. Y., 'Plastiques renforcées: verre ou carbone?', *Usine Nouvelle*, April 1971.
20 PIPER, R. N., Data Recording Instrument Co., Private communication.
21 RICHARDS, A. J., Elliot Automation Ltd, Private communication.
22 ANON, 'Carbon fibres put the squeeze on sound', *New Scientist*, **50**, 747, 138, 1971.
23 LAWES, G., 'Titanium for the aircraft materials prize', *New Scientist*, **45**, 692, 503–505, 1970.
24 ANON, 'CFRP jackets', *Composites*, **2**, 2, 72, 1971.
25 SIMPSON, D. C., 'CFRP jackets', *Composites*, **2**, 3, 137, 1971.
26 ABRAHAMS, M., and DIMMOCK, J., *Mechanical and economical comparisons of reinforced thermoplastics*, Paper 12, Conference on Reinforced Plastics, Plastics Institute, Solihull 1970.

27 ANON, 'Carbon fibre piston rings and PTFE valve shoulders', *The Engineer*, 17 June 1971, 11.
28 CHILDS, R. (RAE), Private communication.
29 DUKES, W. H., *The application of graphite fibre composites to airframe structures*, SAE paper 680316, Society of Automotive Engineers, New York, 1968.
30 RABENHORST, D. W., Johns Hopkins University, Private communication.
31 LOVELL, D. R., *Quality control in carbon fibre manufacture*, Paper 50, International Conference on Carbon Fibres, their Composites and Applications, Plastics Institute, London, 1971.
32 DUKES, R., and GRIFFITHS, D. C., *Marine aspects of carbon fibre and glass fibre/carbon fibre composites*, Paper 28, International Conference on Carbon Fibres, their Composites and Applications, Plastics Institute, London, 1971.
33 PATON, W., 'Carbon fibres in sports equipment', *Composites*, 1, 4, 221–226, 1970.
34 ANON, *New Scientist*, 42, 644, 74, 1969.
35 GOTCH, T. M., *Engineering application of reinforced thermoplastics in rail transport*, Paper 15, Conference on Reinforced Thermoplastics, Plastics Institute, London, 1970.
36 ANON, 'Moulds and press tools', *Composites*, 2, 1, 6, 1971.
37 GILTROW, J. P., 'A design philosophy for carbon fibre reinforced sliding components', *Tribology*, 4, 1, 21–28, 1971.
38 DRAKE, A., *Electro-osmosis*, Data sheet, McCoy-Hill & Partners, Wimborne, Dorset, 1968.
39 STEVENS, D. C., 'An assessment of the oscillating fibre microbalance of Patashnick and Hemenway', *Aerosol Science*, 2, 315–324, 1971.
40 ROBERTS, F., and MORRIS, B., 'Carbon fibres by the ton', *New Scientist*, 51, 759, 68–73, 1971.
41 READ, F. G., 'Discovery makes carbon fibre "cheap as glass"', *The Engineer*, 232, 6008, 27, 1971.
42 ZENDER, G. W., and DEXTER, H. B., *Compressive properties and column efficiencies of metals reinforced on the surface with bonded filaments*, NASA Technical Note TN D-4878, Langley Research Centre, 1968.
43 FISHLOCK, D., 'The route to cheaper power', *Financial Times*, 7 January 1969, 14.
44 SCHREIBER, C. K., 'Polymethylmethacrylate reinforced with carbon fibres', *British Dental Journal*, 130, 1, 29–30, 1971.
45 ANON, 'Carbon fibre reinforced plastics mast', *Plastics and Rubber Weekly*, 258, 7, 1969.
46 FROST, J., 'Carbon fibres', *The Engineer*, 228, 5918, 49, 1969.
47 ANON, *The Engineer*, 230, 5962, 11, 1970.
48 BEDWELL, M., 'Fibre Technology at Morgan Crucible Ltd', *Conference on Fibre Technology in Engineering*, Production Engineering Research Association, Melton Mowbray, 1970.
49 PETERS, D., 'Carbon fibres can spin through generator barrier', *The Engineer*, 230, 5957, 40–41, 1970.
50 WATT, W., and PHILLIPS, L. N., 'Carbon fibres for engineering applications', *Proceedings of the Institution of Mechanical Engineers*, 185, 52/71, 783–806, 1971.
51 JEFFS, E., 'Hyfil was right for fan blade, but time ran out', *The Engineer*, 232, 6000, 81–83, 1971.
52 RABENHORST, D. W., 'Filament flywheel for clean, quiet cars', *Science Journal*, 6, 11, 20, 1970.
53 SOUCH, D. J., and TAYLOR, G., *The development of the pantograph for high-speed collection*, Preprint, Institution of Mechanical Engineers, Railway Division, London, 1971.
54 READ, L. R., National Physical Laboratory, Private communication.
55 MORRIS, J. B., 'Carbon fibres', *Atom*, 144, 269–278, 1968.
56 RUSSELL, J. G., and COLCLOUGH, W. J., *The development of a composite propeller blade with a carbon fibre reinforced plastic spar*, Paper 4, Symposium on the Effect of New Materials on Aircraft Design, Royal Aeronautical Society, London 1971.
57 Ferro Corporation, 'Composites in yacht rigging', *Composites*, 1, 6, 336, 1970.
58 BROGDEN, T., Machine Tool Industry Research Association, Private communication.
59 LUCAS, R., *Exploitation of carbon fibre composites in special purpose plant*, Conference on Practical Applications of Fibre Composites, PERA, Melton Mowbray, September 1971.
60 ROSENBAUM, H. M., Marconi Company Ltd, Private communication.

192

61 DENNISON, H. A., *Design and manufacture in fibre-reinforced plastics*, Conference on Practical Applications of Fibre Composites, PERA, Melton Mowbray, September 1971.
62 FRAY, J., 'A carbon fibre Vulcan airbrake flap', *Journal of Royal Aeronautical Society*, December 1971, 878.
63 The Dunlop Co. Ltd, and Goodyear Tyre & Rubber Co., *Flight International*, **100**, 3276, 995 (1971).
64 G.K.N. Ltd, 'Carbon-fibre-reinforced racing oars', *Composites*, **2**, 4, 208, 1971.

6. Carbon Fibre Brushes for Electrical Machines

J. J. Bates

The material in this chapter has possible commercial significance in the near future. Because of this it has not been possible to be very factual in the presentation of specific data for voltage drops, friction coefficients, wear rates, etc. Such important data are obviously only obtained from lengthy tests, and at this point in time all the results from such tests cannot be made freely available.

6.1 Introduction

Carbon is a very familiar material to electrical engineers in the form of brushes for current collection and commutation. A carbon brush is a composite material formed by mixing carbon powder with some binding agent, followed by subsequent heat treatment. Unlike carbon fibre composites structural strength is not the principal aim. The emphasis is on satisfactory electrical properties, not so much of the brush material itself, but of the brush-to-surface contact. After more than half a century of development such solid brushes have reached some degree of sophistication, and the range of additives to the brushes to fit them for particular applications is very wide.

In the early days of development of electrical machines wire brushes were used, made by bundling metallic wires in a case. Naturally these brushes gave high wear rates, both for themselves and for the surfaces on which they rubbed.

With the arrival of carbon fibres it was perhaps inevitable that the idea was conceived of similarly bundling them together in a case to make a literal 'brush' and it is understood that this was originally suggested by the Scientific Adviser, Director General Ships, MOD (Navy), UK. There would be no point in making a brush this way unless there would be some advantage over a solid brush. Probably the reason for the advantages foreseen lay in the nature of the theories of current collection by a solid brush. One of the most commonly accepted theories maintains that a solid brush has very few contact points at any one time, such contact points representing only a very small part of the nominal brush area and continually changing from point to point on the brush as it moves over the collector surface. With a fibre brush, composed of millions of individual fibres some eight micrometres in diameter there would surely be many more contact points at any one time than in the case of a solid brush and such a brush should be able to handle much higher currents. But it must also be borne in mind that such a brush may not have the natural lubricating properties usually associated with carbon/graphite.

The many-contact-point hypothesis was a logical one, and prototype fibre brushes were produced for evaluation. This chapter is devoted to the research and development that has taken place into the possible uses of such carbon fibre brushes over the last five or six years. Much remains to be done to perfect such brushes (which are not yet commercially available) and explore the ways in which they can contribute to the improved performance of electrical machines.

6.2 The Construction of a Carbon Fibre Brush

To construct a fibre brush the fibres must be bundled together in a metal case with a protruding length which is the working length of the brush. They have to stay there in spite of frictional forces, and satisfactory contact must be made between the case and the fibres. There are several ways of doing this.

6.2.1

As shown in Fig. 6.1, bundles of fibres are placed in two thin-walled channels which can slide together when sufficient fibres are enclosed. The fibres are trimmed flush at the rear end, which can then be capped by:

(a) Copper spraying, forcing copper in between the fibres and adhering to them over a short length.

(b) Pouring in a conducting cement which penetrates somewhat farther than the copper spraying and makes a greater contribution to the structural rigidity of the brush in the sense of retaining some of the shorter broken lengths of fibre which are invariably incorporated during the process of manufacture.

There are two disadvantages in brushes made this way. First it is found that the rectangular shape of the brush is fairly easily lost and the box tends to belly out in time due to the pressure of the fibres. This can cause the brushes to stick in the brush holders. Second, it is difficult to get sufficient pressure on the fibres during manufacture to prevent pull-out of some fibres during use, and the rate of pull-out increases as the packing density gets progressively less.

195

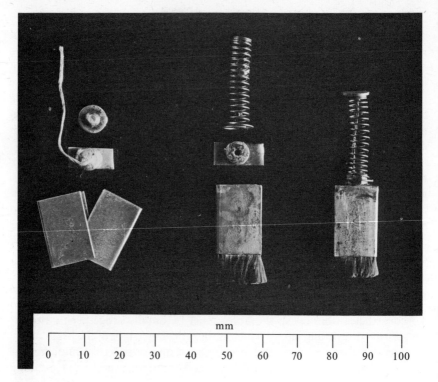

mm

| 0 | 10 | 20 | 30 | 40 | 50 | 60 | 70 | 80 | 90 | 100 |

Fig. 6.1 *Thin-wall fibre brush construction.*

6.2.2

This is illustrated in Fig. 6.2. The aim is to use a thick-walled box and have a much greater packing density for the fibres. A three-sided box A is placed in a jig, and parallel walls are maintained along the length of the jig by loose packing pieces B. A considerable quantity of fibres fills the jig and a top plate C is forced down until it is in position at the top of A, when it is pinned through the jig walls. The brush is then removed, the joints soldered, the fibres trimmed, and an end cap fitted as in 6.2.1.

In many cases the rigidity of the brush can be improved by using metal separators in the boxes as shown in Fig. 6.3. This gives better fibre retention, while at the same time allowing the fibres to spread together over the working length outside the box.

6.2.3

This method seeks to avoid the rather lengthy processes involved in 6.2.1 and 6.2.2 but gives a lower packing density. Fibres are placed in a mould which is then filled with melted paraffin wax, the fibres being put under pressure so

196

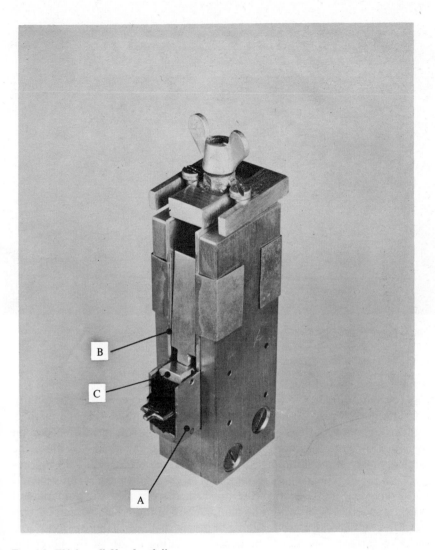

Fig. 6.2 *Thick-wall fibre brush jig.*

that a reasonable packing density is obtained. When the wax solidifies a block of material is obtained which can be cut into brushes of various sizes. These brushes are then placed in their metal cases and the wax removed by heating. The back plate is then fixed on as before.

Figure 6.4 is a sectional photomicrograph of a typical brush made by the process described in 6.2.2 and after running on a slip ring. The dark areas represent the part of the brush which has not been in contact with the ring.

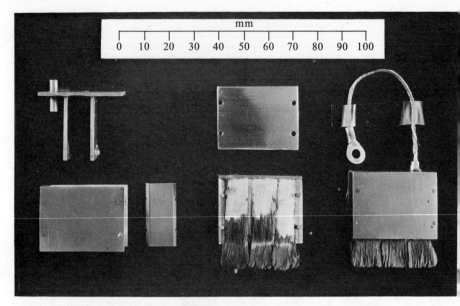

Fig. 6.3 *Thick-wall brush with separators.*

6.3 The Conducting Properties of a Fibre Brush

A fibre brush should have many contact points to the surface on which it is rubbing. Occasionally it is observed that a small part of a brush tends to burn out. This implies that the current becomes restricted to a few fibres for a time. If the rear copper coating or conducting cement is the sole means relied upon to make contact to the fibres then the selective current flow may be along the full length of a limited number of the fibres in the brush and if these fibres burn out the packing density is reduced. The effect may be cumulative, tending to eventual destruction of the brush. If there is a considerable lateral pressure incorporated during the manufacture of the brush, so that current can flow easily across, as well as along the fibres, then even if there is occasional selective current flow along the short protruding length of the fibres the danger of burn-out is less, and even if such burn-out occurs it cannot reduce the packing density of the fibres within the case of the brush, and upon which the brush depends for its rigidity.

The effect of the lateral pressure is shown in Fig. 6.5, pressure being applied to the fibres in the jig shown and resistance being measured between the containing plates. Provided pressures corresponding to the flattening of the curve are applied, the brush should be satisfactory. A brush containing four such sections and thus being approximately 1 cm^2 in section would have a resistance through the brush to the surface contact points of less than 0·01 ohm.

198

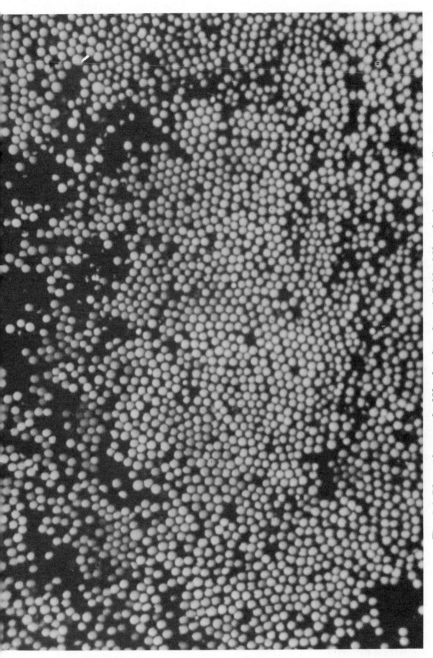

Fig. 6.4 *Photomicrograph of fibre brush surface (individual fibres 8 micrometres diameter).*

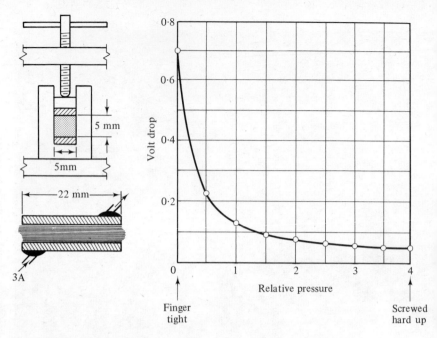

Fig. 6.5 *Effect of lateral pressure on fibre brush resistance.*

The most satisfactory way of assessing the correctness of assembly of a fibre brush is to make two brushes joined together by continuous fibres as shown in Fig. 6.6. The voltage–current curve from end to end of such a double brush is typically as shown in this figure.

To avoid the eventual formation of insulating tarnish films the boxes can be silver plated internally and this is recommended if brushes are to be worked with the maximum possible current density.

6.4 The Sliding Contact Properties of Fibre Brushes

6.4.1 THE BEHAVIOUR ON A COPPER SLIP RING

The most commonly used surface material with solid carbon brushes is copper and its alloys, and it was natural that initial experiments with fibre brushes should also be on such surfaces.

There is a very large number of combinations of the important variables. These are:

(a) Surface speed
(b) Current density
(c) Brush polarity
(d) Common or separate track running

200

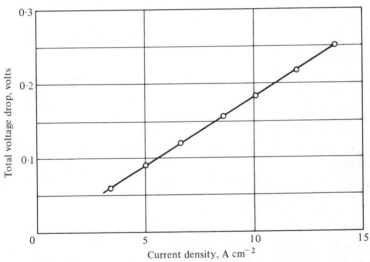

Fig. 6.6 *Double brush for measurement of overall brush resistance.*

(e) Protruding length of fibres
(f) Brush pressure
(g) Brush shape
(h) Surface finish
(i) Surface temperature
(j) Atmospheric conditions.

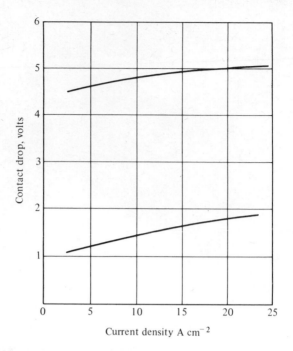

Fig. 6.7 *Typical fibre brush voltage drops.*

Even such parameters as brush shape can be important because of the flexible nature of the brush. If the brush is short in the circumferential direction far more flexing is possible for the majority of the fibres than if the brush is long in the circumferential direction.

Because there are so many variables, all of which have not been fully explored, only a few typical results can be quoted. Figures 6.7 and 6.8 show voltage drops and friction values for separate track running on a copper ring.

Figure 6.7 shows the voltage drops at the two brushes, as a function of current. The voltage drop at the negative brush (brush negative with respect to the surface) is much higher than at the positive brush.

Figure 6.8 shows the coefficients of friction as a function of the surface temperature and here again there is a remarkable difference between the positive brush and the negative brush.

Both voltage drops and friction coefficients are higher than for solid carbon brushes.

The overall conclusion from Figs. 6.7 and 6.8 is that the negative brush can have a high voltage drop but a reasonably low coefficient of friction, while the positive brush can have a low voltage drop and a high coefficient of friction. In either case the combination is such as to make the brushes unattractive

202

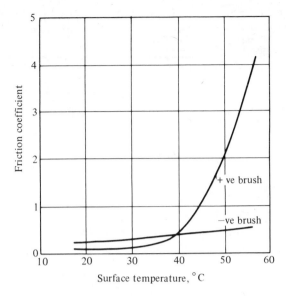

Fig. 6.8 *Effect of surface temperature on friction.*

for current collection of d.c. from copper surfaces compared to conventional carbon brushes.

The surface of the copper under the negative brush became blue-black in colour, but under the positive brush very little marking was evident.

6.4.2 COMMON TRACK RUNNING OF FIBRE BRUSHES

Direct current collection from slip rings means that positive and negative brushes must bear on separate rings and the contact conditions for one polarity can play no part in the behaviour of the brush of opposite polarity. The conditions are different when commutator machines are considered as it is then usually essential for brushes of both polarities to run on a common track.

Tests have been carried out with carbon fibre brushes running on a common track on copper and brass slip rings. It is often very difficult to achieve stable conditions of operation for either the positive brush or the negative brush. The surface may run for a considerable time with little evidence of surface films and with reasonably low voltage drops and coefficients of friction. A dense blue-black film may then appear quite quickly and the voltage drops and the coefficients of friction invariably rise to higher values at both positive and negative brushes. Reasonably stable conditions may then be obtained for a time, but often the voltage drops and coefficients of friction eventually rise again and remarkably high values can be reached. If the current is switched off the dense blue-black track is usually rapidly removed and the

203

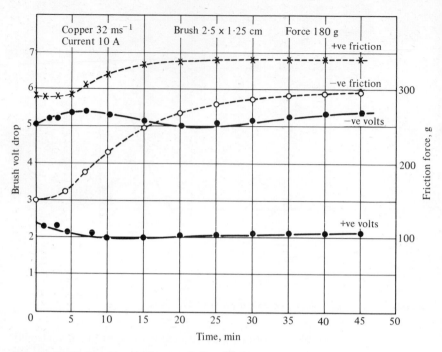

Fig. 6.9 *Variation of brush characteristics with time.*

slip ring assumes a bright copper colour once more. The time required to form the blue-black track is not consistent from day to day even though as far as is practical the conditions of running are the same.

Figure 6.9 shows voltage drops and friction forces as a function of time starting with a clean ring and measured on the rig shown on Fig. 6.10, i.e., with a flat-face brush running on a flat surface and with the brush long in the direction of surface travel. These characteristics are reasonably stable after about an hour's running.

Figure 6.11 shows how dependent the voltage drops and the friction forces are on the surface conditions. From the start up to line AA lubrication was provided by rubbing a conventional EG brush on the previously clean surface. From AA to BB this lubricating brush was removed and it should be noted to what a high value the negative voltage drop rises. At BB the lubricating brush was reapplied and very rapid falls took place in the voltage drops and friction forces.

Many other interesting and peculiar observations have been made. There is evidence of a stick-slip action by the brushes, and there is a strong tendency for the fibres to spread out sideways at right angles to the direction of rotation. This is shown in Fig. 6.12. The fibres appear to catch on the surface in the middle of the brush and be pulled down, bending the outer fibres in the

204

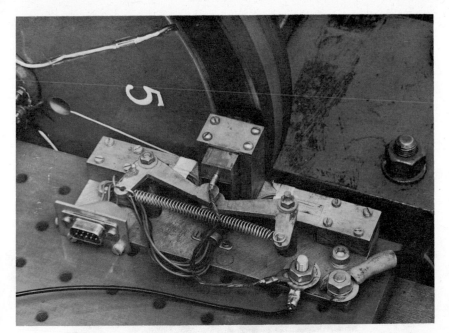

Fig. 6.10 *Surface ring rig for fibre brush tests (note dense surface film).*

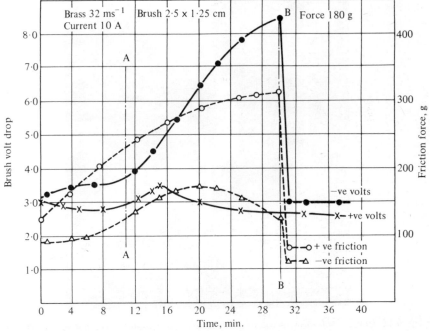

Fig. 6.11 *Effect of lubrication on brush characteristics.*

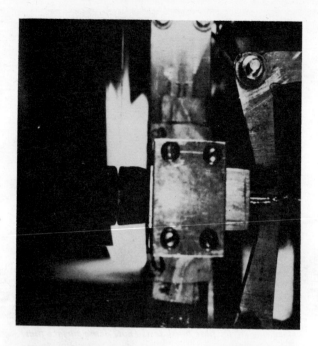

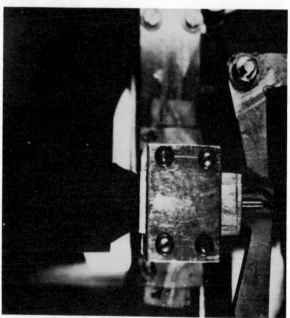

Fig. 6.12 *Unusual sideways spread of fibres for a brush running on the flat surface shown on Fig. 6.10.*

process. There is a tendency to greater spread when the ring is accelerated rapidly from rest instead of being brought up to speed gradually. When the ring is stopped the spread does not disappear until the brush is pulled away from the surface.

The reasons for these peculiar properties of carbon fibre brushes are not as yet established. This is not surprising when it is appreciated that even after half a century of development the behaviour of a solid brush has not really been satisfactorily explained and there are so many more variables with a fibre brush.

Some tentative explanations can be offered for some of the observations. There seems definite evidence that a negative fibre brush establishes a heavy surface film and that a positive brush is equally capable of cleaning off such a film. The film under the negative brush has been analysed and shown to be principally cuprous oxide. The polarity differences imply that conduction is to some extent an ionic process and it is thought that hydroxyl ions from ionized water vapour migrate towards the surface under the negative brush and form a heavy oxide layer. As with common track running, both the coefficients of friction go to high values once the oxide film has formed and this suggests that these high friction effects might be due to sticking of fibres in grooves channelled in the oxide film, remembering that the fibres are individually only some eight micrometres in diameter. The high voltage drops of the brushes, both positive and negative, when running on the oxide film implies that conduction is either through the film or that the voltage has to rise to a sufficiently high value for the film to be broken down electrically. Another fundamental difference between a solid carbon brush and a fibre brush is in the 'memory' of a solid brush. Holm[1] has shown that a solid brush will make repetitive use of contact spots on the surface of a slip ring so that a satisfactory contact is established to some extent by the ability of the solid brush to find and use these contact spots revolution after revolution until these spots are adapted to the particular current collection. With the soft flexible nature of a fibre brush it is very doubtful if such a memory will exist as the fibres will be in continuous perturbation and will be unlikely to make repetitive use of contact spots on the surface.

There is undoubtedly a great deal to be done to understand fully the contact properties of carbon fibre brushes, and from what has been written above it might appear that these brushes are not very useful devices. This is not, however, quite the case. There are certain properties of the fibre brushes that have proved very useful in the field of commutator machines, not as complete replacements for solid brushes, but used in combination with solid brushes to improve the behaviour of the latter under particular conditions. The general philosophy in their useful application has been to recognize the great flexibility of a fibre brush and its superior arc of contact during switching of current between adjacent segments on a commutator. Common track running is a feature of all such promising applications together with a restriction of the current carried so that the tendency of negative brushes to form the unwanted oxide films is successfully opposed by the scrubbing action of the positive brush. A further useful application has arisen from the observed ability of a fibre brush to accept more rapid changes of current than a conventional brush without causing surface damage.

6.4.3 THE CURRENT TRANSFER PROPERTIES OF A FIBRE BRUSH

It is widely recognized that a solid carbon brush only touches the surface on which it is sliding at a few points at any one time, and these points correspond to a very small part of the nominal area of contact. When a solid brush moves from segment to segment as in a commutator machine it is very difficult to maintain a reliable arc of contact in the sense that the brush may take its support entirely from the incoming segment before it has reached the end of its travel across the outgoing segment. This can give a sudden loss of electrical contact with the outgoing segment, or at the very best an uncertain contact during the later stage of the brush movement.

The very flexible nature of a fibre brush means that it fits the surface much more precisely in the presence of small irregularities and has a very reliable arc of contact. The resistance from the brush to a segment varies closely in inverse proportion with the area of brush in contact with the segment. This is shown very simply by Fig. 6.13, which is a measure of the transfer resistance

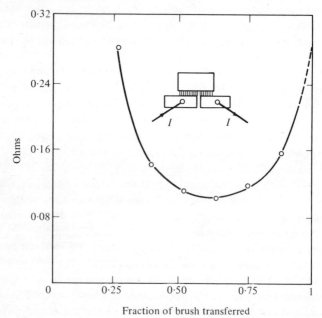

Fig. 6.13 *Transfer resistance of a fibre brush (reproduced by courtesy of IEE).*

208

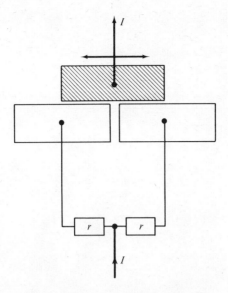

Fig. 6.14 *Principle of current transfer rig (reproduced by courtesy of IEE).*

between two segments as a brush 1 cm × 3 cm in section is slowly moved across the segments.

The transfer properties of solid and fibre brushes have been studied conveniently using the arrangement shown in Fig. 6.14. Here a brush moves between two adjacent segments and the voltage forcing the current transfer can be altered by changing the value of the resistance, r. The brush can be moved relative to the segments by simple low speed reciprocation, by sliding on two rotating semicircular rings or by sliding across two rotating eccentric rings. The latter has been particularly useful in studying the transfer properties of brushes and is illustrated in Fig. 6.15.

Using the rig shown in Fig. 6.15 a comparison has been made of the transfer properties of fibre relative to solid brushes for different values of the resistance r shown on Fig. 6.14. These are shown on Fig. 6.16. The transfer properties of the fibre brushes are far superior to those of solid brushes and correspond closely in all cases to a brush where the brush-to-surface resistance is inversely proportional to the area of contact.

Transfer tests have also been made using chopped carbon fibres incorporated in a solid carbon brush. The results of these show very little improvement over a solid brush.

Some brushes have also been tested using copper-coated carbon fibres. Due to the lower contact resistance of these brushes their transfer properties are again inferior to the plain carbon fibre construction.

Fig. 6.15 *Rig for studying the transfer properties of a fibre brush.*

6.5 The Application of Fibre Brushes to Conventional Commutator Machines

Conventional d.c. machines are designed for satisfactory commutation on the assumption that when a brush is *just* about to leave a segment the inter-pole induced voltage has brought the current flow from the brush to this segment to zero, or a low value. Clearly this can only be the case if the brush-to-segment contact can be relied upon right up to the last moment, and the transfer curves of Fig. 6.16 show that this assumption can be suspect. Particularly at high speeds unreliable contacts are likely and the point of electrical separation of a brush and a segment may be very arbitrary, and sparking will result.

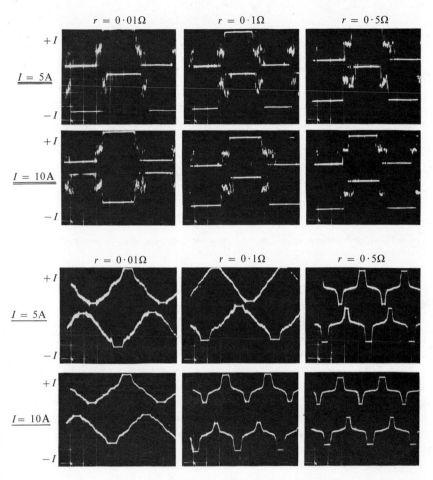

$r = 0.01\Omega$ $r = 0.1\Omega$ $r = 0.5\Omega$

Fig. 6.16 *Transfer properties of fibre and solid brushes—40 reversals/sec. Brush size 0·5 cm × 1 cm and traversing in the 1 cm direction (top set solid brushes, lower set fibre brushes).*

Although the fibre brush has superior arc of contact properties compared with a solid brush there are problems associated with a complete replacement of solid brushes by fibre brushes. The undesirable friction and voltage drop properties associated with common track running and described in section 6.4.2 imply that such brushes would have a limited current carrying capacity and rubbing speed. The most successful way of applying them to a commutator machine is to use them as protective spill-over brushes in parallel with solid brushes as shown in Fig. 6.17. If fibre brushes are used in this way, their action will be as follows:

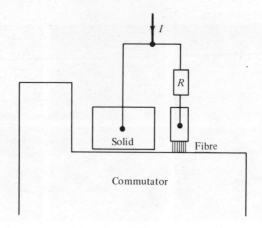

Fig. 6.17 *Fibre brushes used in a protecting mode (reproduced by courtesy of IEE).*

(a) When excessive speed and vibration causes a solid brush occasionally to cease making contact altogether, all the current will pass to the fibre brush which will not lose contact due to its great flexibility.

(b) If a solid brush leaves the outgoing segment earlier than it should, the fibre brush will still be in contact with this segment and the final stage of commutation will be completed by the fibre brush.

(c) A fibre brush has a higher transverse resistance than a solid brush, and can resistively commutate more current than a solid brush. When a combination of solid and fibre brushes resistively commutate, the solid brush will initially carry most of the current. But as its area of contact gets less its current density, and the brush-to-segment voltage will rise. Eventually the density and the voltage drop will reach arcing values, but when this tends to happen the current will automatically move more to the fibre brush and the final stage of the resistance commutation will be done by the fibre brush. In other words the fibre brush clamps the voltage that can build up between the solid brush and the surface and prevents arc formation.

Figure 6.18 shows the relative solid and fibre brush currents measured on a d.c. machine with a small fibre brush alongside a solid brush and with a small resistance in series with the fibre brush to limit the current it can be asked to carry. The use of the parallel running fibre brush gave a considerable increase in the current that could be handled by the machine without sparking. The average current carried by the fibre brushes is seen to be very low and no heavy oxide film formation was observed. An interesting observation is the peaking of the fibre brush currents at *slot* frequency not bar frequency, probably as the commutation takes place between the last conductor in a slot and the first conductor in the next slot, as it was observed that the machine tested had a selective bar marking sometimes found in d.c. machines.

212

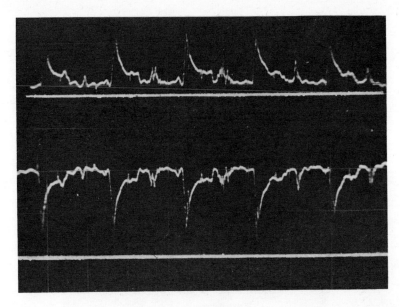

Fig. 6.18 *Relative solid and fibre brush currents on a d.c. machine (top trace fibre; bottom trace solid) (reproduced by courtesy of IEE).*

6.6 The Application of Carbon Fibre Brushes to a TAC Machine

Considerable development has been going on in recent years on a class of machines where commutation is assisted by semiconductor devices. This is known as thyristor assisted commutation (TAC). The principle of action of these machines is shown on Fig. 6.19. The single commutator of a conventional machine is replaced by two commutators each containing a number of active and inactive sections. The brushes are also split into so called part-brushes, and part-brush pairs are connected together through two back-to-back thyristors.

The principle of operation is that when brush B2 is just fully on the active segment C_{22} thyristor T_2 is fired, thyristor T_1 being already conducting. The interpole induced voltage in the commutating loop formed by the coil 12 and the thyristors must now reduce the current to zero in brush A1 before this brush reaches the end of segment C_{11}. The particular point at which the current is brought to zero is not important provided it occurs before the brush A1 leaves the segment to an extent that would make its contact uncertain. This arrangement ensures that brushes never interrupt current and the commutation limit of such machines is much greater than for a conventional d.c. machine and it is possible to make d.c. machines in hitherto impossible power-speed combinations.

213

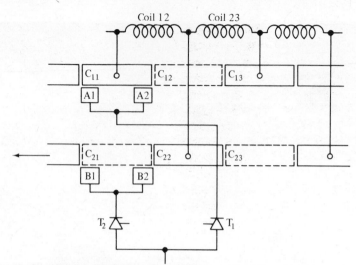

Fig. 6.19 *Principle of action of a TAC machine.*

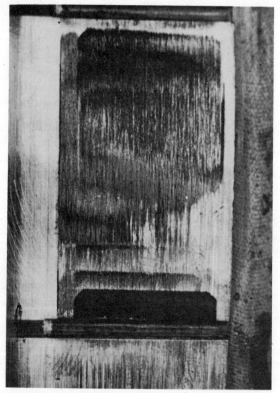

Fig. 6.20 *Surface conditions on TAC machine without fibre brushes (reproduced by courtesy of IEE).*

214

Although the brushes in these machines do not have to break current, they are working in a peculiar manner, akin to a pulse of current at high frequency to a brush running on a slip ring. It was found in the development of these machines that solid brushes would not accept such current pulses without marking the surface, and no grade of solid brush could be found that did not mark the surface.

The tendency of solid brushes to mark the surface under such conditions was overcome by putting fibre brushes in parallel with solid brushes to protect them against the shock loading. It was found that the fibre brushes would accept such pulses without causing surface damage, but their higher voltage drop meant that when the pulse had passed and the solid brushes were able to behave satisfactorily the majority of the current would pass to the solid brushes. The average current carried by the fibre brushes was once again limited to a low value, and common track running was possible without the formation of heavy oxide films and the consequent problems.

Figure 6.20 shows the surface conditions on a TAC machine without the protective action of the fibre brushes, and Fig. 6.21 is the corresponding case when a (smaller) fibre brush is run in parallel with the solid brush, which now forms a smooth even skin with no surface damage. Figure 6.22 shows a number of commutation cycles with the fibre brush currents a small proportion of the total thyristor current when acting in this protecting manner.

The arrangement of Fig. 6.19 will work equally well if the thyristors are replaced by diodes except that now current will start to flow through brush

Fig. 6.21 *Surface conditions on TAC machine with the protective action of fibre brushes (l.h. track fibre, r.h. track solid).*

215

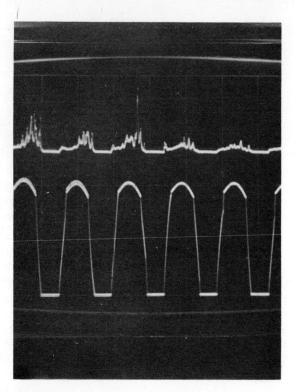

Fig. 6.22 *Fibre and thyristor currents on a TAC machine.*

B2 as soon as it comes into contact with the edge of the active segment. Due to the high coil voltages in these machines with few armature tappings this always produces severe entering edge damage when solid brushes alone are used.

If the solid brushes are, however, led into the active segments by fibre brushes, the surface damage is drastically reduced and in some machines it is possible to use the simpler arrangement of diodes instead of thyristors. Figure 6.23 shows how fibre brushes are so used and how the current initially through the fibre brushes transfers mostly to the solid brush once this has become adequately established on the surface. This is a very good illustration of the clamping action of a fibre brush to limit the voltage between a solid brush and the surface until the solid brush contact makes, at a relatively low brush-to-surface voltage and consequently without damage.

6.7 The Design of Brush Boxes for use with Carbon Fibre Brushes

There are some fundamentally different requirements between the brush boxes for use with carbon fibre brushes and those for use with solid carbon

216

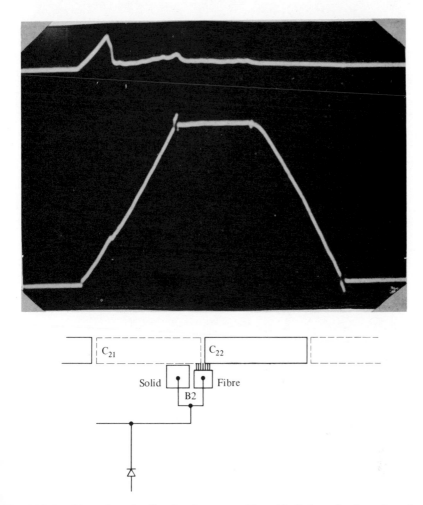

Fig. 6.23 *Lead-in action of a fibre brush on a machine with diode assisted commutation (upper trace fibre brush current; lower trace diode current) (reproduced by courtesy of IEE).*

brushes. This is because of the metallic nature of the case used in the manufacture of the carbon fibre brush. If the brush box is also of metal and there is no lubricant, which there cannot be because of the danger of its finding its way down the fibres and onto the contact surface, it is found that the brushes can stick in the boxes, or at the very best the pressure on the brush is very indeterminate.

Another undesirable effect is the tendency of current to bypass the brush pigtail and enter through the walls of the brush. This is shown on Fig. 6.24

217

Fig. 6.24 *Burning on walls of fibre brush.*

where marking can be seen on the outside of the brush, indicating that current has entered here at the path of least resistance to the surface. Currents entering the brush through the box walls tend to cause sparking, roughening, and make the tendency of the brushes to stick in the boxes worse. Currents also tend to flow in through the pressure fingers. These effects are of course absent in the case of solid carbon brushes, where there is as good a lubrication between the brush and the walls of the brush boxes as there is between the brush and the surface.

To overcome these defects the brush boxes have been constructed of an insulating material and lined with PTFE faced metal sheet, and the top of the brush has also been insulated. A typical brush and box are thus as shown in Fig. 6.25.

218

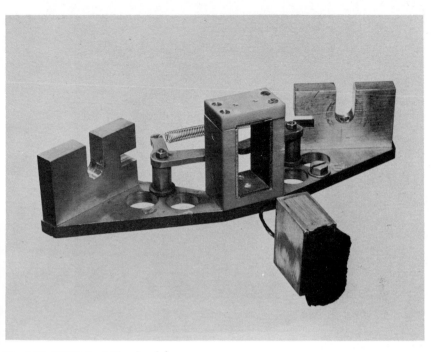

Fig. 6.25 *PTFE lined fibre brush box.*

6.8 Plated Carbon Fibre Brushes

Much of the work on carbon fibre brushes was initiated in the hope that such brushes, by virtue of their many contact points, would be capable of passing very high currents and would be particularly applicable to the heavy current, low voltage, homopolar d.c. machines currently under development and using superconducting field coils.

As the early work showed that this would not be the case for plain fibres due to the high voltage drops that could occur, a considerable amount of development work has been carried out using metal plated fibres, each individual fibre receiving a metal coating before being bundled together to form a brush.

Initial results with plated fibres showed high friction forces and wear rates, though low voltage drops. The latter were typically of the order of 0·2 V for the negative brush and 0·8 V for the positive brush at current densities even as high as 900 kA m^{-2}. However, if the brushes are run in inert atmospheres the wear rates are considerably reduced and figures quoted for a run of 2000 hours are 1·5 mm for the negative brush and 6 mm for the positive brush.

The extent to which the carbon fibre plays a part in current transfer or in the lubrication of such brushes is not very clear at present. It is possible that

219

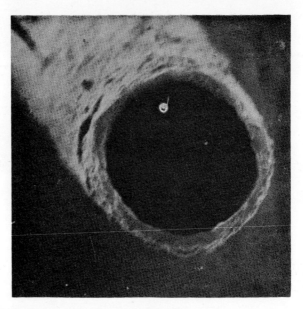

Fig. 6.26 *Metal-plated carbon fibre (reproduced by courtesy of IEE).*

such brushes are essentially similar to metal wire brushes with the principal difference that it is possible to use the strength of a fibre to support the metal to provide a very flexible wire for incorporation in a brush. The flexibility may then allow the plated fibres to bend up and down over surface irregularities to an extent that would not be possible with an all-metal wire of the same diameter. This would make the friction and wear of a plated fibre brush much less than that of an equivalent wire brush, and it is possible that this is the reason for the success obtained with the plated fibre brushes.

Plated fibre brushes will not be successful on commutator machines, where a reasonably high contact resistance is necessary to assist current transfer from segment to segment. Figure 6.26 shows the plating on the outside of an 8 μm carbon fibre used in the brush construction, and Fig. 6.27 shows the appearance of a completed brush.

As well as plating the fibres, a considerable number of tests have been carried out using plain fibres running on various plated surfaces in an attempt to reduce the voltage drops and make the brushes suitable for very heavy current use but these tests were generally unsuccessful. Where voltage drops were reduced, high friction forces were invariably obtained.

6.9 The Continuing Development of Fibre Brushes

It must be emphasized that carbon fibre brushes are very much in the stage of prototype development. But they are showing themselves to be extremely

220

Fig. 6.27 *Plated fibre prototype brush (reproduced by courtesy of IEE).*

promising in the field of commutator machines. Work is currently proceeding to evaluate their protective action on a wide variety of machines, always used in conjunction with solid brushes and attempting to avoid the formation of the harmful heavy oxide films. As well as applying them to existing machines by simply replacing one solid brush on a brush arm by a fibre brush, special rigs are in use to evaluate the way a solid brush is protected by a fibre brush and whether the protective action is solely at the trailing edge of a solid brush or also occurs partly towards the centre or even possibly towards the leading edge of the solid brush.

It will be noticed that no values have been quoted for wear rates. Information is still being gathered on this, but very long running periods are required. Used incorrectly, wear rates can be high, but provided the heavy oxide films can be avoided, and high surface temperatures are avoided, a typical wear rate could be as low as 1 mm in 1000 h of running at surface speeds of some $25 \, \text{m s}^{-1}$.

It is possible that fibre brushes will show up to advantage on a.c. machines also. With the constantly changing polarity of a brush with a.c. use there is not the tendency to set up such heavy oxide films, as the film laid down when the brush is negative is cleaned off when it changes to positive. The higher transverse resistance of the fibre brush should also lower circulating currents when used in a.c. commutator machines.

221

6.10 Acknowledgements

The author is indebted to Morganite Carbon, International Research and Development, the UK Admiralty Materials Laboratory, the National Research and Development Corporation, and colleagues at the Royal Military College of Science for their cooperation and assistance.

References

1 HOLM, RAGNOR, *Electrical Contacts, Theory and Applications*, Springer-Verlag, New York, 1967.
2 BATES, J. J., 'Thyristor Assisted Commutation in Electrical Machines', *Proc. IEE*, June 1968.
3 BATES, J. J., STANWAY, J., and SANSUM, R. F., 'Contact Problems in Machines using Thyristor Assisted Commutation', *Proc. IEE*, Feb. 1970.
4 BATES, J. J., 'Diode Assisted Commutation', *Proc. IEE*, May 1970.
5 BATES, J. J., and POWELL, R., 'Relative Current Switching Properties of Carbon Fibre Brushes and Solid Brushes and their Implications for Commutator Machines', *Proc. IEE*, April 1971.
6 PETERS, D., 'Carbon Fibres Could Sweep Away Commutator Troubles', *The Engineer*, Nov. 1970.
7 McNAB, I. R., and WILKIN, G. A., 'Carbon Fibre Brushes for Superconducting Machines', *Journal IEE*, Jan. 1972.
8 Morganite Carbon Ltd, *Carbon and its Uses*, 1971.

7. Quality Control of Carbon Fibre Materials

W. G. Cook and A. G. Downhill

7.1 Introduction

7.1.1 MEANING OF QUALITY

It is now generally recognized that the quality of a product is the effective combination of the quality of the basic design supported by quality of conformance to the design. At the same time it is less generally appreciated that the two aspects of quality are inter-related and that features of a design may automatically render quality of conformance difficult or even impossible to establish. This applies both to dimensional parameters and also freedom from defects likely to fail the product.

An important corollary of this aspect is the fact that some products are released for a specified life and have then to be re-inspected, perhaps repaired and re-certified and released for further use on a progressive scale. This makes it paramount that appropriate parameters, e.g., freedom from cracks, corrosion, joint integrity, etc., are capable of economic detection and assessment.

Since many of these features affecting inspectability and maintainability are inherent in the basic design, quality control starts when research begins on new concepts and data are generated which provide the designer with the fundamentals of materials, processes, and manufacturing capabilities on which to base his new specifications.

7.1.2 DESIGN REVIEW

The quality task at this stage is to identify quality problems at the earliest possible stage and initiate action either to eliminate the problem by redesign (often a very simple task if tackled early) or initiate quality research where the feature is fundamentally involved with the design. Needless to say, quality problems eliminated at this stage are the least costly of all since the remedy needs draughting time only and no raw material or hardware is yet involved. The quality research may be to develop existing techniques (say non-destructive testing) where the application demands capabilities beyond the

223

known state of the art. Alternatively the need may arise to develop entirely new or novel techniques. It is now recognized that research into quality problems arising in the new designs can require as much lead time as the basic project.

7.1.3 EXPERIMENTAL MANUFACTURE

Unfortunately with the best will in the world it is not always possible to identify all these problems at the design stage and the activity continues in the experimental or development manufacturing phase when the first samples of hardware based on the new concept can first be handled and assessed. The elimination of quality problems from the design at this stage is more expensive since raw material has been ordered and worked upon or alternatively there is less lead time for planning research.

Fortunately at this time the designer is sometimes in the same dilemma and a redesign may provide the required opportunity to incorporate quality thinking into the redesigned product. Once the first products are available the development phase of evaluation by testing begins and there also begins a further round of design changes to remedy the undesirable features revealed by the testing. Here Quality Control plays a key role since the designer can only recognize success on the basis of accurate knowledge of what he tested, the testing environment and an equally accurate report on the results of testing.

7.1.4 STANDARDIZATION AND PRODUCTION

The production configuration is standardized and the manufacturing organization then has the responsibility of delivering to the customer, with Quality Control playing its more conventionally accepted role. At the same time it must be appreciated that the development programme, however extensive, is never capable of sorting out all problems and Production Quality Control must continue reporting process incompatibilities and design problems. When production initially starts this may represent a high level of activity arising from new manufacturing sources and learning curves.

7.1.5 SERVICE

The product delivered to the customer is exposed to usage and as experience builds up there must be an effective feedback from service to ensure that the failures experienced and the lessons learnt result in effective and timely improvement to the original specification. At the same time the customer is given a service of spares, repair schemes, overhaul procedures, technical guidance, and product support. Quality Engineering again is a major factor in many of these activities. The customers' complaints are handled effectively and speedily and he is kept advised of remedial actions taken to prevent recurrence of trouble. Quality Engineering plays an integrated part in this.

Although the defects arising in operational use are usually predominantly design weaknesses, there is also a proportion of instances where delivered items show departures from specification.

These range over the field of activity, manufacturing, configuration control, packaging, correctness of paperwork, and so on. In serious cases it may be necessary to answer the question 'How many more like this?' and have an effective system for taking adequate recovery action.

Thus Quality Engineering is a 'cradle to grave' activity and the data generated by Quality at all phases play a key role in maintaining standards and leading to a continually improved product. The basic activities and flow of information from the design stage to standardization is shown in Fig. 7.1

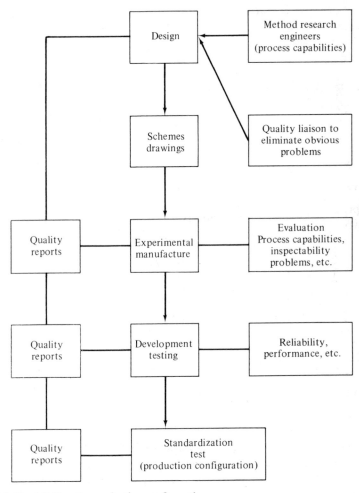

Fig. 7.1 *Establishing the production configuration.*

and the flow of information in the Production and Service phases is a logical extension of this.

7.2 Quality Engineering on New Projects

7.2.1 QUALITY AS PART OF THE MANAGEMENT TEAM

The increasing involvement of Quality Engineering on new projects has revealed not only the need for personnel with adequate knowledge and authority as well as adaptability to new techniques but also the need to appreciate the very important human and management aspects involved. In a process-orientated manufacturing programme the closest understanding and co-operation is essential between the Design, Production Engineering, Laboratory, and Quality team involved.

In the early stages of new projects the designer is receiving information, data and claims from the enthusiastic specialists pioneering the new technology. These specialists have a difficult task trying to persuade designers to discard established materials and procedures and to adopt new and untried ones. The evidence is not always factual and often includes over-optimistic extrapolation of facts in the light of foreseen or desired improvements.

Sometimes, on the other hand, the designer grasps at some new but undeveloped potentiality in his eagerness to get the best design. This makes it essential that claims and forecasts are effectively analysed to avoid the design specification demanding requirements which cannot be achieved on a production scale.

The role of Quality Engineering is to provide data to establish that the creative efforts of the designers in generating the specification and the performance of the product engineer in meeting the specification are sufficiently in harmony and that the company does not become committed to an impossible or uneconomic task.

Pure research workers and technical specialists often have little grounding in Quality Control and this is often reflected in the management of the programme. The development of Quality Engineering techniques and equipment needs time, manpower, facilities, and components on which to work, often for long periods. Unless Quality Engineering is actively integrated into the programme from the start, these needs may be ignored and it often requires major trouble to make management appreciate these facts.

7.2.2 QUALITY ENGINEERING APPROACH

At the same time new materials such as carbon fibres and the associated technologies introduce a host of novel technical terms and jargon, often with a halo of magic around them. In the urge to get away from the conventional approaches it is often not easy to convince everyone that despite the novel

226

elements the whole situation is amenable to classical Quality Engineering and that no new fundamental principles are needed.

Whatever the terminology or complexity of the process the prime need is to start with a clear and adequate specification. Manufacturing endeavours to produce to the specification and Quality Control to check the achievement. Non-conformance to specification is analysed and fed back for corrective action by Production Engineering, Design, or both and the situation continuously monitored. The accepted products (conforming or cleared by Design as having acceptable deviations) are then subjected to functional testing to evaluate the design. This is usually done under controlled conditions designed to establish that the product is capable of meeting its intended purposes over the full envelope of environmental conditions with appropriate extensions where these are needed to give adequate assurance. Any modifications shown to be necessary are then incorporated and the cycle of events repeated until a satisfactory product is achieved and can be standardized.

In the case of carbon fibre products the procedures are more complex than normal. This primarily springs from the fact that although the designer may know the final configuration he requires, the process of arriving at this through a building-up process of laminating and (possibly) moulding, raises problems with the design details of the intermediate shapes and forms. In order to achieve the stress-carrying properties by the use of preferred fibre orientation it is obvious that these intermediate stages have to be carefully controlled. In consequence the stages of manufacture are more critically related to the final product than is common with most manufacturing techniques in more conventional materials. This inevitably leads to a great deal of evaluation of shapes, lay-ups, moulding techniques, and processing which may involve many cycles of 'cut and try' evaluation before workable design configuration can be achieved. All this work has to be done under controlled conditions so that when a workable configuration is established the process can be repeated accurately.

7.2.3 UNLEARN 'METAL' THINKING

As with glass fibre laminates the transition from metal to carbon fibre composites raises special problems. In the first transition, it is very difficult to unlearn 'metal' thinking. The behaviour and modes of degradation of metal components are well known and normally catered for. Although a great deal has still to be learned, it is well established that the behaviour of composites shows very great differences and these need to be understood and embodied in the design. This is most clearly brought to light when a product normally made in metal is made in composite material without a basic redesign. The problems of learning curves in new ventures are well recognized but 'unlearning' is perhaps less appreciated and takes longer.

227

A further serious problem which arises at this stage is that of dealing with material defects and the systems needed for developing acceptance standards. These standards, often called 'Quality Acceptance Standards' although legally an extension of the drawing, aim to cover the controi of non-dimensional attributes such as porosity, inclusions, cracks, voids, and similar undesirable features to which all materials are liable.

The problems associated with these troubles are sometimes chronic in the early phases and many of the prototype components have to be released for test and evaluation with numerous unresolved problems in this respect. In some cases, however, small pilot plants and laboratory conditions can produce over-optimistic results in relation to what happens on a production scale, when the real spread of variability is encountered. Carbon fibre composites show both sides of this problem. Where the means of detecting these defects are themselves limited or ineffectual to varying degrees, the remedy is to resort to destructive cut-ups and sectioning.

There are two inter-linked problems—firstly to have the ability to detect the 'defects' and secondly to assess whether they matter. In practice the latter stage has to be pre-judged, at least initially, and efforts directed to detecting the things which are thought to matter.

Subsequent experience may reverse or support the judgement and will assuredly add other defects to the list. The process of establishing which apparent defects really matter, which are innocuous and what the threshold of acceptance is, both for individual attributes as well as combinations of attributes, represents a serious problem. This is still difficult in dealing with more conventional materials and is even more so on carbon fibre composites.

The objective is a written acceptance standard with clearly defined limitations. In practice with most materials and especially with carbon fibre composites, verbal definitions are inadequate and the standards must include illustrations in the form of photographs and line drawings, indicating acceptable and rejectable levels of defect. At the same time the aim should be to set a standard enabling inspection to pass judgement on the bulk of the work it has to deal with and accept the fact that the remainder may need special consideration, especially in the early stages.

The difficulties are serious since too high a standard can lead to excessive rejections, high scrap costs, and disruption of programme. On the other hand too low a standard means release of work incapable of meeting requirements, leading to wasted efforts, scrap, and liabilities which can be far more expensive than the value of the parts. Ideally the standard represents the optimum compromise between the process capability and the economics of the situation within the actual functional requirements. A major factor is often the lack of firm knowledge of the functional behaviour of the defects. The subject does not lend itself readily to mathematical treatment or stress analysis and there are virtually no scientific aids in assessing defects in any way comparable with those developed for finding the defects. There is often in consequence a wide range of opinion. Designers and stress engineers often press for complete elimination of these types of defects, sometimes regardless of their significance, and at the other end of the scale there are output-orientated managers who press for maximum relaxation.

The worst situation is where the responsible authorities abdicate and leave the whole problem to the man on the shop floor—usually the inspector. This can usually be recognized by such phrases as 'shall be free from harmful defects' or 'shall not contain excessive inclusions' and the like. The only effective solution is testing under the appropriate environmental conditions on a statistically significant basis. This must be done so that defects are exposed to cycles of stress, temperature, and other degrading conditions. The most direct way is to use the product in its actual application but this is not always possible or economically feasible by testing to long lives or special (say) altitude conditions in the early stages of a programme before hardware is available. Also, testing a part to destruction when it is part of a complete mechanism may be extremely expensive if the result is major damage to the assembly. The remedy is to use various forms of rig testing. These are very flexible and whilst it is often difficult or impossible to reproduce perfectly a particular environment of stress patterns, temperature conditions and gradients, and cyclic changes in these patterns, a very effective compromise can often be reached. In some cases by the careful use of temperature factors, overstressing, etc., essential data in accelerated testing can be achieved which would be impracticable in the true environment. An additional feature which is often more easily handled in a rig environment, is the ability to study modes of degradation over controlled periods. Similarly, once basic data on performance have been established the efficacy of repairs, salvage schemes, dimensional deviations, and design changes can be speedily assessed and cleared.

The main factors to be considered in establishing Quality Acceptance Standards are shown in Fig. 7.2. This indicates the complex nature of the task and serves to stress the importance of the fact that the standards should be developed as an integral part of the development programme so that as many facets as possible are defined in readiness for the Production phases.

7.2.5 PRODUCTION PHASE

The transition to Production is often a painful procedure. It is sometimes imagined that the defects and problems experienced in the development phase will somehow go away in the production phase. This is probably induced by the efforts put into the new tools, facilities, etc., which are needed for quantity production. In reality the results are very mixed, the improved facilities do eliminate certain defects encountered in development but often

229

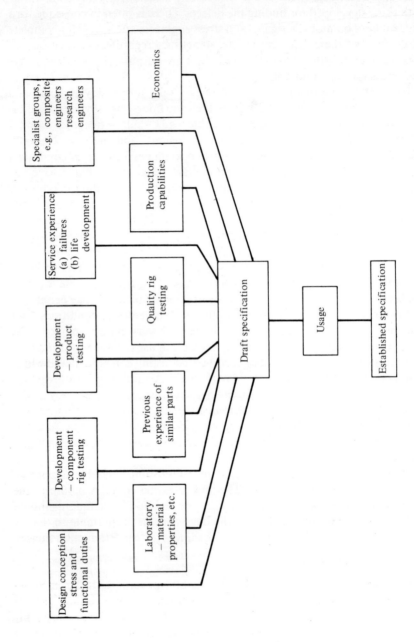

Fig. 7.2 *Sources of information to establish quality acceptance standards.*

they do not. On the other hand they frequently bring with them a range of new troubles. The situation has to be treated on its merits and the move from an Experimental Quality Acceptance Standard to a Production version has to be phased and managed very carefully. Where Production plans fail to achieve plan targets and it is necessary to resort to assistance from the Development Facilities, this problem can be most acute.

7.3 Quality Control—Carbon Fibre Products

7.3.1 GENERAL

Objectives. Whilst the Quality Control objectives of any manufacturing system are basically the same in that there must be adequate manpower and facilities effectively integrated into the total management system to achieve the specified objectives, there are special aspects of carbon fibre technology which call for special emphasis and treatment. As a member of the family of fibre composite materials, carbon fibres share many of the problems of that family with, however, a number of problems specific to carbon fibres.

The basic ability inherent in composite fibre materials of being able to start with a thin sheet of pre-impregnated material and build up by a series of processes to produce a consolidated structure with the critical fibres suitably orientated in a cured resin matrix is at once the great strength and potential weakness of the process. While the process is inherently flexible, the integrity of the component is fundamentally dependent on a number of critical features which cannot be effectively checked non-destructively after completion of manufacture. The integrity of these built-up composites depends on ability of the fibres to carry the stresses and this can be drastically reduced if broken, kinked, or mal-positioned fibres exist. Built-up structures depend on the integrity of adhesive joints and bonded interfaces. Other factors are resin content (this may vary from core to skin) porosity, degree of cure, inclusions, omission (or addition) of laminates, and internal cracks and voids.

In the complex manufacturing cycle for a die-moulded carbon fibre composite the sequence of events may include the following—cutting of laminates with controlled size and fibre orientation, build of pre forms in sub-assemblies and assemblies, pre-cure of assemblies to achieve the correct mouldable state, hot die moulding, and post-cure of moulding to achieve ultimate cure and stability. All these can be critical and require detail study in each phase.

By their nature carbon fibres lose two advantages found with glass fibre composites which show a degree of transparency or at least translucency which enables visual inspection (either by incident or transmitted light) to be of some value. This facility is absent with opaque carbon fibres. While destructive tests in the form of resin burn out can be used on glass composites to establish the correct lay-up of laminates, this is not always practicable with carbon fibres and destructive sectioning is both time-consuming and much less informative than carrying out a three-dimensional survey. In view of these limitations and the problem of effective evaluation in the finished state it is essential to base the Quality plan on process control. This means that the manufacturing processes and controls must be fully and clearly defined and rigidly adhered to under conditions of adequate cleanliness. At the same time it is of paramount importance that management and personnel at all levels are fully appreciative of the situation and maintain the disciplines from personal conviction rather than coercion.

General facilities. The layout and planning of the general facilities is based on the following considerations:

(a) Storage of incoming raw materials—refrigerators and bonded stores
(b) Cleanliness by control of airborne contaminants from external sources, unsealed dust-generating floors, dust-generating manufacturing processes, etc.
(c) Operator-generated contaminants can be minimized by special consideration of protective clothing, laundering, washrooms, and segregated facilities for meals, refreshments, and smoking. These often raise secondary problems but it should be appreciated that there are also health factors of major importance.

The equipment, both general and special-to-purpose, must be planned with provision for adequate controls of critical parameters, e.g., temperature, times, and pressures, if necessary on a continuous recording system.

Personnel. It is of paramount importance that the prime functions concerned with the production and certification of a product of acceptable quality are adequately staffed and working as a team within the management structure. This concerns Laboratory, Production Engineers, and Quality personnel (both Inspection and Quality Engineers), bearing in mind the needs for investigations and trouble shooting as well as the routine clearance of components.

7.3.2 RAW MATERIAL

Types. Fundamentally the raw materials are carbon fibre and resin. However, apart from filament winding this is not the most convenient form and some preparation of the ingredients is required before receipt by the manufacturer. While for glass fibre the fibres are often unimpregnated and procured in the form of a laminate, either undirectional fibre or woven cloth, it is normal to procure carbon fibre in a resin pre-impregnated form either as undirectional laminate or in some instances as chopped fibre where random directional laminate is required for certain mouldings.

232

Procurement control. In principle there is little difference between the control required for procuring carbon fibre 'prepreg' and forgings or high duty alloys. In either case Quality assurance must depend to a large degree on the process control of the supplies to a declared material specification. This is confirmed by a sampling receiving inspection which will include laboratory tests to determine the material properties. However, since the main aspects of the material relate to its final cured state the test cannot be performed on the material in its supplied form. It is normal to design and manufacture test pieces which are cured and the physical properties measured by appropriate tests.

In the case of bonding resins the test pieces would take the form of specially prepared joints. It is important that all such tests should be mutually agreed between supplier and customer and will often constitute part of the suppliers' acceptance testing.

Visual inspection on receipt plays a greater part than for metals to identify such faults as fibre mal-alignment, lack of cleanliness, and inclusions. This inspection can also confirm adequate packaging, lack of damage in transit, and the existence of documenting evidence such as date stamp, shelf life, etc. The former is important because inadequate protection during transit whether by intention or damage could allow contamination or exposure to a detrimental environment which may not be detected by the normal tests on receipt.

Storage. An important aspect of resinous materials is that of shelf life and deterioration is dependent on storage time, handling, contamination, and environment in respect of humidity and temperature. Furthermore, deterioration which seriously affects the usability of the material is not necessarily evident like the corrosion of metals, or in some cases not even detectable by laboratory tests. To discover failings in raw material storage at the time of final inspection of the product cannot be entertained. It is therefore imperative that the requirements for storage, handling, shelf life, mixes, etc., must be detailed and rigidly controlled. Accurate provisioning and a system of 'first in, first out' is essential. It is worth noting that in some ways ageing is a form of precure and process errors during the pre-cure stage of moulding probably have more effect on the final fibre orientation and porosity than any other parameter. For most pre-impregnated material long-term storage can only be accomplished in refrigerators and it is necessary that surveys are carried out to ensure temperatures are sufficiently low. Most commercial cold storage is around 0°C whereas effective storage of resins should be at a temperature lower than − 10°C. In removing material from a refrigerator for issue, certain precautions must be taken such as to allow sufficient time for the material to reach room temperature before opening the hermetically sealed container to avoid condensation. In view of the possibility of identical appearance of

233

differing materials identification is vital. This cannot normally be done by marking the material and one is reliant on marking the packaging. In the case of 'prepreg' a convenient identification method is to use colour-coded backing paper which remains as part of the material up to the point of usage.

7.3.3 MANUFACTURING CONTROL

Process evaluation. Since the designer is concerned with the orientation of the fibres in the finished structure he must necessarily be concerned in the orientation prior to moulding or curing. The design must specify therefore not only final product requirements but also the numbers, shapes, and positions of the individual laminates. This may be relatively simple in the laying-up of multiple layers of unidirectional fibres or very complex when considering die-moulded components with different direction of fibre in different layers or material with different resin/carbon ratios within the composite. Accomplishment of the design must involve a knowledge of the movement of the fibres during moulding. The result is likely to be an empirical compromise established by a series of trials and may only approximate to the theoretical, but possibly impractical, design.

This evaluation needed to arrive at a combination of design and manufacturing specifications must be accepted as an integral part of the programme in advance of the normal product development phase necessary to evaluate the product suitability in use. The exercise involves a much closer Design/Quality/Production Engineer relationship than normally experienced in metal technology, to minimize the timescale of the change feedback loop.

The whole exercise must be supported by a wide range of destructive, non-destructive, dimensional, and functional testing to establish both the integrity of the product and the process capabilities. An important 'spin off' of this programme is the early identification of the critical attributes on which monitoring of the process can be based.

On satisfactory completion of the evaluation exercise, the design and manufacturing specification and quality acceptance standards can be formally standardized and the manufacturing controls and special safeguards defined. Subsequent changes to these requirements must be carefully controlled and considered in the light of need to repeat the evaluation process —a costly and time-consuming exercise.

Process control. The necessity of adequate process control in lieu of a policy of 'Inspect and Reject' has been emphasized earlier. Having proved that a product satisfies the original intent both by evaluation and later by product development, Production management must control the manufacture to produce consistently a similar product. It is the task of Quality in its inspectional role to monitor the process and this can be divided into three

234

functions. First, the confirmation of conditions before work is commenced. This can include the more obvious aspects like checking surface preparation, e.g., water break tests, confirmation of material controls being observed, and checking the contents of assembly kits particularly in the case when numerous laminates are involved. In the case of bonded joints where adhesive thickness is important for joint integrity it may be necessary to confirm joint gaps. This may range from simple measurement to complete dry assembly including, sometimes, the full curing cycle, using one of the commercially available materials developed for this purpose. Selective assembly may also be necessary.

The second function is the monitoring of the process parameters by patrol inspection. An important aspect here is the advantage of recording instruments coupled to the equipment instrumentation. Not only does this ease the operator's task in maintaining control but also minimizes the inspection task in confirming it. It is obvious that all equipment instrumentation must be included in the overall scheme for calibration of measurement equipment. However, in some instances the actual equipment must be subject to calibration in aspects not normally measured, e.g., oven temperature distribution searches.

The third function is the measurement of attributes in the actual manufactured product either complete or part complete as appropriate or in some cases prepared test pieces. This will obviously include dimensional, NDT, and sample destructive testing, more details of which will be found in subsequent sections. Whilst it is accepted that in some instances the inspection may be important in avoiding the existence of a particular critical defect in the supplied product its other important role is to confirm that the measured attributes fall within the predetermined process capability thus inferring that the process is under control. In some cases considerable effort is put into the measurement of an unimportant attribute because the evaluation indicated it to be a sensitive monitor of an important aspect of the process, a factor often forgotten in times of cost reduction.

Documentation. The keystone of process control is an efficient system of documentation. It is advantageous to develop a completely integrated system to satisfy the requirements of the various company departments. In this way apart from the avoidance of excessive complexity, inspection is automatically involved and does not have to be on the alert for work to be performed with the danger of missed inspections and ensuing problems.

The essence of process control is a detailed *plan* to satisfy the requirements of the *specification*, measurement and record of the *performance* in meeting the plan, a *standard* for assessment of performance, and *quality assurance* of conformity. It is convenient to consider a system of documentation in relation to these categories. Whilst the various documents have been referred to in the main text where applicable it is thought advantageous to summarize them in this paragraph.

(a) Specification	Drawing
	Laboratory Process Specification
	Material Specification
	Quality Specification
(b) Plan	Manufacturing procedure (operation list)
	Inspection procedure (including dimensional, NDT, and destructive testing)
(c) Performance Record	Process control card or route card
(d) Acceptance Standard	Drawing tolerances (dimensional)
	Quality Acceptance Standards (non-dimensional)
(e) Quality Assurance	Inspection records
	Inspection marking on the component (where applicable)

Rework, salvage, and repair. While these are somewhat similar functions and often confused it is worthwhile recalling the differences normally associated with these terms. Rework is the recovery of non-conforming products using the original manufacturing techniques thus rendering them identical to the standard product. Salvage on the other hand is the recovery of non-conformance by a different method from the original which renders the product acceptable for use. Repair is the recovery of damaged components.

Rework is common practice in metal work particularly when the error is in the 'metal on' condition. In the manufacture of carbon fibre (or glass fibre) products the manufacturing techniques normally involve unrepeatable processes (moulding, curing, etc.) and hence rework is less common and salvage is necessary.

The fragility of many components inevitably leads to damage due to handling (or mishandling) in spite of the precautions taken and hence repair must be considered even for a new product. In some cases this may result from part of the process, e.g., delamination caused by machining.

Repair and salvage do necessitate two aspects which need special attention. Firstly the techniques used must be detailed, evaluated, and approved in the same way as the normal manufacture and must be subject to the same disciplines of control. Secondly it must be recognized that this work will be required on production items and hence it is imperative that all techniques must be cleared during the development phase to establish acceptability in the intended use.

Maximum freedom must be given to the local Quality department to include previously cleared salvage and repair schemes as required. In some

236

instances their inclusion can be an integral part of the production process to be carried out on the products in the batch as necessary, e.g., sealing of fibres.

The records accumulated during any such work must be considered with the main manufacturing records for final acceptance of the product and where applicable stored for traceability.

7.3.4 DIMENSIONAL INSPECTION

Final dimensional inspection of fully cured carbon fibre composites can raise metrology problems due to the complexity of the configuration but, in general, raises no serious difficulties intrinsically attributable to the use of carbon fibres. However, during the various process stages problems can arise which merit discussion.

The individual contoured laminates used in hot die moulding may cover a range of shapes with preferred fibre orientation. These shapes may be cut from lofted templates in the early stage of development or cut in large numbers by contoured cutters when throughput justifies the tooling. The product is, of course, a soft tacky material and dimensional control is assured by optical projection on a sampling basis. Similarly fibre orientation must be secured basically by datum marks on the cutters and material guides, and overchecks carried out on a sampling basis.

When the laminates have been 'laid-up' the assembly can be very critical in that it must be inserted into the die in the correct state. The pliable nature of the product at this stage makes it essential to support the assembly in properly designed cavities to avoid distortion and slip. These factors can drastically affect the resin flow and fibre movement during the moulding stage. In the case of large mouldings the final curing of the component may be effected as a post-curing operation in order to improve utilization of the dies. Again support may be necessary to avoid distortion. Dimensional problems are also encountered in complex structures where the flexibility of intermediate shapes and repeated heating as assemblies are built up can cause dimensional deviations which are difficult to eliminate. This can necessitate a planned programme of recording numerous parameters as the assembly is built up in order to pinpoint the source of trouble. As with some problems in sheet metals the degrees of distortion due to shrinkage, coefficients of expansion and the like can lead to deliberate policies of making details incorrect by a controlled amount to ensure that they are correct in the finished state.

7.3.5 NON-DESTRUCTIVE TESTING

Typical defects. Before discussing the scope and limitations of available techniques it is pertinent to deal with the typical defects likely to be encountered.

237

There are numerous crack-like defects. These range from surface crazing to internal cracks which may be interlaminar, translaminar, or transfibrous (see Fig. 7.3). In this class can be included delamination and splitting. The

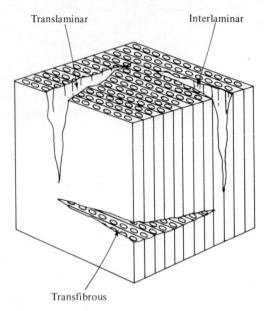

Fig. 7.3 *Definition of cracks: Interlaminar—between laminates; Translaminar—across laminates; Transfibrous—fibre fracture.*

significance of these defects is often difficult to assess. Fatigue testing of laminar specimens in first flap certainly shows a markedly different behaviour from metals. Metal specimens resonated in first flap maintain frequency as fatigue damage accumulates and a drop of one cycle per second is positive indication of the initiation of cracking. This then propagates rapidly and failure is imminent with most metals. With carbon composites, as with glass, the frequency may drop as much as 1 per cent and then continue for a long period without failure. As the degradation proceeds the frequency drops in small decrements and eventually the damping capacity becomes so great that the test has to be discontinued. Thus surface or internal cracks in carbon fibre composites cannot necessarily always be regarded as being areas of high stress concentration likely to cause premature failures.

Fibre orientation problems include broken fibres, kinked fibres, fibre 'wash', fibre distortion, fibre packing, and mal-aligned fibres. All these can be critical and affect the strength of the product. They can be on the surface or inside the component, and can arise at various stages of the manufacturing process.

238

Resin defects include resin starvation, resin richness, porosity, incorrectly cured resin, excess resin as well as incorrect identity of resins and associated compounds, hardness, plasticizers, etc. The problems may be internal or on the surface.

If the raw material is tacky, this can lead to the pick-up of numerous foreign bodies from the manufacturing process, its environment or from operators. These may be metallic or non-metallic including pieces of backing paper from packing sources.

Unbonded joints and joints of low strength can exist for a variety of reasons arising from lack of cleanliness, poor joint preparation, incorrect gaps and glue line thickness, excessive entrapment of air, poor resin mix, inadvertent presence of parting agents, etc. These can be critical and in such cases components are liable to fail at very short lives.

Non-destructive testing techniques. Carbon fibre composites present difficult problems since many of the standard techniques developed for metals have either no application or only limited capabilities.

Surface inspection for defects presents serious problems. The completely non-magnetic characteristics of resin and fibres make the use of classical magnetic methods impossible. Ostensibly penetrant methods are suitable but in practice the surface finish, minute crazing, and the 'wicking' effect of exposed fibres leads to spurious indications and can create hazards for subsequent operations, e.g., bonding, so that penetrants are little used.

Visual inspection is also very limited due to opaque nature of the product and there are no techniques available for enhancing the contrast between defect and background comparable with etching, electropolishing, etc., which can form the basis of very critical inspection for metals.

Some non-metallics, e.g., ceramics, are amenable to inspection for surface cracks by the use of electrostatic techniques where particles of fine powder cause a build-up at cracks due to the high electrostatic potential at the edges of the cracks. The basic conductivity of carbon fibre excludes the use of the process as a means of surface inspection.

Although carbon fibres are electrical conductors and show some response to eddy currents, it has not proved practicable to detect cracks and the like even on simple shapes so that it seems a remote possibility for complex shapes. At the same time the use of eddy currents to detect resin-rich surface areas seems feasible for components of simple shape. Unfortunately these problems tend to arise in radii and corners of fairly complex geometry or below the surface. In the case of metals there exists a range of facilities capable of identifying and sorting some alloys and the heat treatment conditions of alloys. These techniques are based on eddy-current measurements of resistance and permeability, thermo-electric measurements, tribo-electric measurements, and similar physical properties.

239

In the case of resins such techniques are very limited. Measurement of electrical resistance, hardness, capacitance, and dielectric properties and spectroscopic analysis in the ultraviolet and infrared range have all proved insufficiently discriminating to form the basis of a practical shop floor application.

For internal inspection the use of various transmission techniques gives some assistance although again the scope is limited. The use of low voltage radiography and beryllium windows renders visible internal cracks in the plane of the beam, with the usual limitations of radiography in detecting tight cracks or cracks not orientated in the beam. Unfortunately the contrast is inadequate to resolve fibres and kinked, broken, or swirling fibres cannot be shown up directly. However, the lay of fibres can be deduced by the use of trace fibres in denser material, e.g., platinum wire or even lead glass, interspaced in the structure. These show up on x-ray although care is needed in interpreting the results and the assumption that the trace elements follow the fibres may be incorrect unless due precautions are taken. Radiography is of very limited value in the inspection of bonds and joints where voids tend to be at right angles to the beam.

Ultrasonic testing in the lower frequency range tends to be complementary to radiography. It has a maximum sensitivity to defects at right angles to the beam and is capable of detecting unbond conditions, interlaminar voids and cavities. Transmission techniques are more commonly used, although pulse echo techniques are also feasible. Ultrasonics is of course at its best with simple shapes and becomes less effective as the geometry becomes complex, although this can be overcome to some degree, by the use of water-coupled probes, wheel probes, or full immersion testing. For other than simple shapes the probe programming may be complex in order to scan the component effectively.

Ultrasonic techniques are also useful in the inspection of bonded joints and honeycomb structures. Single probe pulse echo systems are often feasible and a number of ultrasonic instruments designed to sense the damping characteristics of the surface layers are available. The Fokker Bond Tester is the one most widely used. Thermal scanning techniques are also available. These depend on the ability to detect a transient heat pattern on the surface caused by a build-up of heat in unbonded area and heat transfer in bonded area. Although elaborate infra-red cameras have been used experimentally it is more acceptable to use heat-sensitive papers or fluids to detect these patterns when heat is applied to the surface. These techniques are dependent on thermal characteristics leading to an adequate build-up of heat pattern. They are consequently of little use with solid mouldings but may be of use with honeycombs.

For skins and joints of suitable geometry, resonance, differential pressures, and sonic testing (e.g., tapping) can all have their application and there is often a direct read across from established glass fibre and metal honeycomb practices. Acoustic emission is a very promising technique for evaluation of composites, the emission being either in the audible range or ultrasonic and a great deal of development is going on.

General conclusions. The limitations of the available non-destructive testing techniques coupled with the complete absence of some techniques for assessing certain critical parameters emphasize the vital importance of process control and the need to supplement the controls by adequate programmes of functional and destructive testing where these can be meaningful.

7.3.6 DESTRUCTIVE TESTING

Sample testing. Destructive testing plays an important role in the process of evaluating and standardizing new processes and in the case of carbon fibres this can carry on into the manufacturing process as a means of monitoring those characteristics which cannot be assessed non-destructively. Thus destructive testing can be used for the following purposes:

(a) Cut up for assessment of parameters otherwise impossible to assess— e.g., kinked fibres
(b) Destructive functional tests, e.g., overload to failure, to assess margins of safety.

Both processes may be used to give confidence in the process controls and the rest of the batch or group produced at the same time. Such processes are expensive and time-consuming if the sample is to be statistically significant and can lead to very difficult decisions when the samples reveal discrepancies and by implication complex batch traceability systems are necessary. For these reasons these techniques should be approached with considerable caution.

Test pieces. Where the product, say an assembly, is large and expensive and it is necessary to establish that the critical parameters are correct, test pieces can be used in the absence of valid non-destructive testing in the product. Thus a test piece using the same materials and resins processed identically with the product can give evidence by destructive testing that the product is satisfactory in some respects. There are, however, obvious pitfalls and considerable controls are necessary to ensure that the test is a valid representation of the product. Apart from technical difficulties when pressures are applied by vacuum bags, etc., it is often difficult to ensure complete similarity, and failures of test pieces due to (say) substandard cleaning or different handling, can raise serious problems. Nevertheless with good technical controls they can give evidence of many critical parameters which are difficult or impossible to assess in any other way.

241

The use of functional testing as a means of acceptance of a component is an important facility when dealing with an uninspectable characteristic. This is not peculiar to carbon fibre composites but can occur with metals.

Where a critical component, say a pressure vessel, cannot be adequately cleared by non-destructive testing or by the use of test pieces, recourse is had to functional testing by exposing the vessel to enough cycles of stress to validate the structure. On the basis of considerable testing it is possible, in some cases, to devise a test which detects weak or faulty joints in a reasonable number of cycles of stress without sacrificing more than a fraction of a percentage of the actual fatigue life of a good component. The use of repeated cycles of stress is important and this is far more searching than a single pressure test which may prove completely misleading if it fails to test the component adequately or even harmful if it causes damage without immediate failure. Such tests need to be the subject of careful technical assessment.

7.4 Conclusions

The Quality Control of carbon fibre composite materials can be based on conventional policies. Nevertheless the special problems of the manufacturing techniques and the limitations of the available non-destructive testing processes highlight the vital importance of effective process control in depth and the need for an integrated approach with Quality aspects clearly understood and catered for from the beginning.

This approach needs to be continued through into the manufacturing cycle both in establishing the techniques and maintaining them in order to achieve a product of consistent and acceptable quality. Product assurance can be further supported by a range of activities such as destructive testing and functional testing which may be used to a greater extent than on most conventional materials. For management to rely on inspection at the end of the line is demonstrably ineffectual.

7.5 Acknowledgements

The authors wish to acknowledge the assistance given by numerous colleagues within the Rolls-Royce organization. The permission of the management of Rolls-Royce (1971) Ltd to publish the information is also acknowledged and it must be stressed that the opinions expressed are those of the authors and do not necessarily represent the official policy of the Company.

8. The Future of Carbon Fibres

Marcus Langley

It would be unwise to forecast the future of carbon fibres or to extrapolate present trends into the sphere of science fiction. Man-made fibres, 'whiskers', glasses, ceramics, and synthetic polymeric materials are all developing exponentially and will undoubtedly continue to do so, challenging metals in many branches of engineering and at the same time competing with each other.

The preceding chapters have indicated some of the lines of development now being followed. Some of these may prove to be unprofitable dead ends, while others may lead to as yet unexpected applications. Much work is being done behind the screens of military and commercial security but it is of interest to review the indications given by the various authors in the earlier chapters as well as in the published papers referred to in the text.

Starting with the carbon fibres themselves, only one diameter (7·5–8·5 μm) has so far been made commercially in Great Britain and therefore everything else, the independent uses of the fibres and the types of matrices, relates to that diameter. But it was reported (p. 14) that strong, high modulus fibres have been made with diameters as large as 30 μm using pitch as the precursor instead of the PAN which has so far been preferred. Other precursors such as cellulose fibres have also been tried with promising results and it may be that these other materials will be used in the production of fibres which are technically useful and commercially viable.

It appears that the properties of the fibres may be modified by subsequent treatments. Surface etching has been found to improve the bond with the matrix and thus increase the interlaminar shear strength. High temperature steam treatment and irradiation with neutron dosage improves the fibre strength and modulus. Such laboratory investigations will continue and may give better commercial products if the industrial processes are not too costly.

Although useful matrix materials have been found amongst the wide range of synthetic resins—the epoxies, polyesters, phenolics, and polyimides—this field is wide open for further developments, not only in the improvement of these particular types but for research into new ones. However, as was pointed

243

out in chapter 1, other quite different matrix materials are being investigated. Some metals, but not all, show promise and work is being done on ceramics, glasses, graphite, and even on concrete, but there are still technical difficulties to be resolved.

The whole range of techniques of using CFC is still too new for the time element in environmental degradation to have produced evidence equivalent to the corrosion, fatigue, and creep of metals found in service. The fibres themselves appear to be relatively inert and permanent under normal conditions and to have good fatigue properties. The polymer matrices have been generally known and used for other purposes over a number of years. Nevertheless this is not enough if the use is a critical one as in aircraft structures or nuclear power plants, but something may be learned from accelerated environmental testing. In aerospace work, the general principle followed both in Great Britain and the USA has been to start with the smaller, less critical units and only to progress towards major components as the smaller ones are found to be satisfactory. This policy will probably be continued over a period of some years before the first all-CFC aircraft structures are put into regular service. (In a twelve-month period 1971–2 six CFRP trim tabs were tested in flight on RAF Jet Provosts and completed 1500 hours service. The conditions included hailstorms and flying speeds throughout the subsonic range up to high altitudes and aerobatics. The only problems came from handling damage and these resulted in the development of repair techniques. The tests were to lead on to the production of a spoiler and fin as a trial fitment for the Anglo-French Jaguar in the British and French air forces.)

The methods of handling and working glass fibre reinforced materials already developed were quickly adapted for use with carbon fibres and proved very suitable owing to the similarity of the matrices. In fact mixtures of glass and carbon fibres have been used to advantage in certain applications. There will be developments in the automatic control of press and moulding processes to reduce the time factor and adjust it to the stage of curing. Similarly one may expect tape-laying and filament winding to be mechanized. Pultrusion will be further developed for the manufacture of standard and special sections. Tapes, sheets, and woven materials, sized and preimpregnated, already appear in the manufacturers' catalogues. Aligned chopped fibre prepreg may have many uses, particularly where double curvature is required.

While such semi-finished materials will undoubtedly have many uses there is a strong tendency for structural design to be expressed in tailored components of as large dimensions as possible to minimize the use of joints, about which much more remains to be known if they are to be efficient in weight.

Heath[1] suggests that CFC structures may come to resemble a development of the Barnes Wallis geodetics of the Vickers Wellington bombers with a closely spaced network of criss-cross fibres held in place and made impervious by the resin matrix. He also suggests that the neglected framework concept attributed to Michell[2] might find practical expression in the use of CFRP.

As yet the use of CFC in internal combustion engines has not shown great promise because of the temperature limitations of the polymer matrix materials. Something better than the polyimides will have to be found and it may be that one must await the successful development of ceramic and other matrices before carbon fibres can be used in the hot zones of combustion chambers. Nevertheless there is considerable promise in the use of CFC in gears, bearings, and shafts. The bodywork and chassis of automobiles may eventually incorporate carbon fibres or carbon fibre/glass composites at suitable points.

A method of selecting materials and of gauging their suitability for particular applications was given in chapter 5. The relatively short list of possible uses shown there will undoubtedly be extended in the future. In mechanisms the properties which have the greatest appeal are those which will reduce the inertia of moving parts, increase their stiffness, and improve their temperature stability. In some cases advantage may be taken of the low coefficient of friction, the rate of wear, the absence of plastic yield and creep, and of the good fatigue properties of CFC when correctly applied. In other cases the non-magnetic properties may be desirable, whether in the large structure of a minesweeper, in weaponry or in fine instruments.

Some indications of electrical applications for carbon fibres were given in chapters 5 and 6. Of these their use in the brushes of electrical machines is particularly interesting. Many other uses will undoubtedly appear.

In the chemical industry the advantages of carbon over glass fibres have been noted. The present limitation here, as in other fields, is in the polymer matrices and their resistance to chemical attack and high temperatures.

A big extension of carbon fibre usage may be expected in medical engineering and equipment, especially when the purposes are more important than the cost. One may foresee big developments in the field of artificial limbs and harnesses and also in surgical implants when the problems of the compatibility of the matrix materials with the body fluids have been solved.

The full potential of CFC will not be realized until the variability of the materials has been narrowed down to within closer limits by means of quality control, as shown in chapter 7. At present there are many points along the line from the raw precursors to the finished products where variations can creep in.

The properties of the fibres themselves are sensitive to controls in their manufacture and they may be used in matrices which are liable to variations. The storage time and conditions of the matrix materials can be very important. Standards are being set up and official specifications are available.[3,4] Even when these standards are observed, however, there are still many random

245

variations which can occur due to lay-up and moulding techniques which may lead to poor quality products.

Inspection is difficult in an opaque material which does not respond to radiography and this has led to the use of thermal scanning, sonics, and ultrasonics.

One may not have complete confidence in the quality and integrity of the finished product without the destructive testing of samples and the non-destructive testing of the final components.

8.1 Economic Factors

Carbon fibres were undoubtedly oversold in the early days, pushed along by ill-informed publicity in the daily press where extravagant claims were head-lined by non-technical journalists. It can justly be said that the fibres them-selves have lived up to the original assessment made by the Farnborough inventors and that the disappointments have been associated with inadequate matrices, that is to say inadequate for certain applications which were then considered. Now, however, work is going forward more steadily as proto-types are developed and satisfactory service histories appear.

At first, engineers engaged on commercial and industrial projects were hesitant because of the high price of the fibre, then quoted at £135 per pound. This limited serious work to defence and aerospace applications where efficiency was more important than cost. Furthermore, at that time the material was only made in lengths of one metre, the process being 'labour intensive', but now continuous length production on automatic plant is general and the price has fallen considerably. This trend will continue as the market expands and competition becomes more active.

An analogy may be found here with aluminium, which was offered in the eighteen-sixties as a laboratory curiosity at £2·50 per pound but which now fetches about 16p per pound, the difference being even greater if one allows for the change in the value of money over the last century. Aluminium is now a serious competitor with steel even though its price on a weight-for-weight basis is higher.

The important point here is not the price per pound or kilogramme but rather the properties that one is buying for a given sum of money. If one material is twice as stiff as another, other things being equal, it may carry up to twice the price if stiffness is what one wants. Further, as in the heddle frame (p. 169), a higher price may be justified by an improvement in performance of the product, especially when the more expensive material is only used at particular points where it has an advantage.

It is always difficult to assess the value of weight saving in commercially engineered products and in some cases it may even be a disadvantage, as for example in divers' boots and in drop hammers. In other cases weight and inertia have to be considered as in rapidly oscillating movements. The importance of weight saving can be shown most simply in aerospace vehicles. In a vertical-take-off aircraft, every pound or kilogramme to be lifted off needs not less than the equivalent of one and a quarter times that weight in thrust and this sets up a vicious spiral of weight increments in terms of engine weight and fuel.

In conventional fixed wing aircraft, the advantage of saving weight in the structure and equipment can be shown in the following simple way. Given that a passenger and his luggage weighs 100 kilogrammes, that his fare is of the order of £10 per flying hour, and that the aircraft has a utilization of 2500 hours per year with a useful life of 10 years, one can show that the revenue potential of one kilogramme of tare weight saved is:

$$\frac{\text{£10 per hour} \times 2500 \text{ hours per year} \times 10 \text{ years}}{100 \text{ kilogrammes}} = \text{£2500 per kilogramme}$$

The whole of such a return will not be achieved in practical airline operations since aircraft do not always fly at maximum capacity and there may be interest charges on borrowed capital but even if one takes a quarter of this figure it still justifies the use of a more expensive structural material.

An evaluation of this kind is easier to make with commercial aviation than with other forms of transport, since these may have many more factors in the equation. Nevertheless ship, railway, and road transport operators are very alive to the advantages of light tare weights if these can be achieved profitably.

From a British point of view, carbon fibre developments may prove particularly important since the equivalent metals are mostly made from imported ores, with heavy freightage. Such materials as the PAN precursor from which the fibres are made can be entirely home produced and are relatively inexpensive. When the manufacturing processes have been made largely automatic and needing only machine tending, and when the capital cost has been written off, the major factor in the price will be the cost of electricity to heat the furnaces. One may then expect the manufacture to be centred on those places where electric power is cheapest or where waste heat, as perhaps at a nuclear power station, is available at a high enough temperature.

For greater stiffness, higher furnace temperatures are needed in the final stage of manufacture and therefore the costs will be higher. It may be that the demand for this form of the material will not be as great as for the other grades and that in fact cheaper materials of inferior properties will still have a considerable commercial market. In this context one should note that carbon fibres with different properties can be made from rayon, tar, and perhaps other precursors and that prices have not stabilized themselves.

247

8.2 Design Philosophy

The definite outline of a design philosophy appropriate to the use of carbon fibre composites is gradually emerging, but slowly as more engineers begin to use them in widely different branches of industry.

The starting point must be the realization that one is not dealing with one material but with a group having a wide range of properties. Within the limits of the group, the design engineer with sufficient knowledge may write down his own 'recipe' or design his material to suit his purpose. He may vary the mixtures of fibres and matrix across the section of a one-piece unit or he may build up sandwiches with other materials to meet the local stresses, and he may vary the orientation of the fibres to suit. In structural work the engineer used to ferroconcrete or wood may find this way of thinking easier than will one used to steel frameworks.

There are no short-cut, rule-of-thumb methods yet available and perhaps there never will be. Every design problem will have to be approached from first principles and with imagination. It will be remembered how the first railway coaches and motor cars derived their outlines and details from the horse-drawn carriages which they supplanted and how it took many years for designers to break away from traditional habits of thought. Naval architects continued to design masts into ships long after sails had disappeared.

Originality in design is slow to develop, due perhaps to limited knowledge, caution or resistance to change, but more often it is hampered by the absence of imagination.

Success in the use of Carbon Fibre Composites will come first to those who are both knowledgeable and wide minded.

References

1 HEATH, W. G., 'Carbon Fibre Composites—Promises and Problems', *AGARD Symposium, Conference Proceedings*, Sept. 1972, Toulouse.
2 COX, H. L., *The Design of Structures of Least Weight*, Pergamon Press, 1965.
3 *Provisional Specification for Carbon Fibres*. NM/538/539/541/546 D MAT, Ministry of Defence, London, February 1971.
4 *Provisional Specification, Resin Pre-impregnated Carbon Fibre Tape and Sheet*. Issue 3, NM 547, D MAT/AVIATION, Ministry of Defence, London, July 1971.

248

Appendix. Physical Properties of Carbon Fibres and Composites

THE DATA GIVEN IN THIS APPENDIX ARE FOR USE IN PRELIMINARY DESIGN STUDIES, THERE BEING CONSIDERABLE VARIABILITY IN THE MATERIALS AT ALL STAGES OF MANUFACTURE. SPECIMENS AND PROTOTYPE PRODUCTS SHOULD BE SUBJECTED TO SAMPLE AND ENVIRONMENTAL TESTING AND THEREAFTER PRODUCTION SHOULD BE UNDER STRICT QUALITY CONTROL.

Two sets of information are given:

(a) Extracts from the official British specifications NM/538/539/541/546 and 547, D/MAT/Aviation, Procurement Executive, Ministry of Defence (Aviation Supply).

(b) Extracts from manufacturers' data sheets, information published in various papers and journals, etc.

Revised information is likely to be issued from time to time as the technology develops, and the user should therefore attempt to keep up to date. The official document and the manufacturers' data sheets should be referred to for further information.

A.1 Official Specifications

The minimum values are all based on a nominal 10 000 filaments per tow. The designations used are:

S   following the type number signifies that the fibre has been surface finished to improve the adhesion of the fibres to a resin matrix. The effect of this treatment should appear in a considerable improvement in the values of ILSS (interlaminar shear strength).

C   signifies Continuous Filament, which is fibre supplied in multi-filament form without twist in continuous length not less than 200 metres, the filaments being substantially continuous throughout the length.

LS   signifies Long Staple supplied in multi-filament tow form with twist not more than 4 turns per metre of continuous lengths not less than 1 metre in which assemblies are of substantially parallel-laid tows.

249

Fibre designation Type	1SC/ 10 000	1SLS/ 10 000	2C/ 10 000	2LS/ 10 000	2SC/ 10 000	2SLS/ 10 000	3C/ 10 000	3LS/ 10 000
Tests on single fibres Diameter range μm	7·0–10·0	6·0–9·0	7·0–10·0	6·0–9·0	7·0–10·0	6·0–9·0	7·5–10·5	6·5–9·5
Minimum tensile strength MN m^{-2}	1700	1700	2250	2400	2250	2400	1900	1900
Tensile modulus range GN m^{-2}	345–415	345–415	220–275	240–290	220–275	240–290	180–220	190–240
Tests of fibre-resin composites Minimum tensile strength at 60% fibre content by volume MN m^{-2}	920	920	1200	1300	1200	1300	1030	1030
Minimum tensile modulus at 60% fibre content by volume GN m^{-2}	187	187	115	130	115	130	97	103
Minimum inter-laminar shear strength at 55–65% fibre content by volume MN m^{-2}	48	48	23	23	69	69	55	55

A.2 Manufacturer's Values—Fibres and Composites

The manufacturer whose values are quoted below guarantees to provide material not inferior to that required to meet the official Specification.

Other matrix materials might be expected to give different values for composites.

Fibre properties from current production—short length

This table shows the analyses of quality control test results for fibre tested during December 1970, January and February 1971.

Property	Units	Fibre Type 1-S Mean	SD*	2-S Mean	SD*	3-S Mean	SD*
Tensile Modulus	$GN\,m^{-2}$	442	34·5	255	15·2	221	15·2
Ultimate Tensile Strength (5 cm gauge length)	$MN\,m^{-2}$	1725	276	2970	290	2450	318
Diameter	μm	7·9	0·1	8·45	0·07	8·5	0·27
Weight per unit length	tex	970	20	960	40	930	30
	$g\,m^{-1}$	0·97	0·02	0·96	0·04	0·93	0·03
Density	$g\,cm^{-3}$	1.98		1·78		1·78	
ILSS (828/NMA/BDMA 60% v/v fibre 5:1 Span:Depth)	$MN\,m^{-2}$	50·4	8·3	79·4	2·8	77·3	2·8

*SD = Standard Deviation

Typical properties of unidirectional 60 per cent v/v composite bars. Shell 'Epikote' 828/MNA/ BDMA resin system

Property	Units	Fibre Type 1-S Typical Mean	C of V	2-S Typical Mean	C of V	3-S Typical Mean	C of V
Composite Tensile Modulus	$GN\,m^{-2}$	210	4	140	4	106	6
Composite Ultimate Tensile Strength	$MN\,m^{-2}$	1200	10	1600	8	1300	9
Composite Short Beam Shear Strength	$MN\,m^{-2}$	68	14	81	14	85	—
Fibre Tensile (Young's) Modulus	$GN\,m^{-2}$	350		234		207	
Fibre Ultimate Tensile Strength	$MN\,m^{-2}$	2000		2670		2170	
Fibre Diameter	μm	7·7	2	8·0	1	8·3	2
Fibre Density	$g\,cm^{-3}$	1·90	2	1·77	3	1·65	3
Mass per unit length of Tow	tex ($g\,km^{-1}$)	810	6	860	6	900	6

Notes:

(a) These data apply to long-length fibre.

(b) 'C of V'—Coefficient of Variation between test batches $= \dfrac{\text{standard deviation}}{\text{mean value}} \times 100$.

(c) 'Mean'—Mean value = Average value measured for 10 kg (22 lb) test batch of fibre.

(d) Fibre tensile modulus and strength values calculated from composite properties.

251

A.3 Resistance of Polymer Matrix Materials to Attack by Chemicals and Solvents

Chemical engineers and those engaged in oil refinery work have been using glass reinforced plastics for many years in the form of tanks, pipes, and ducting, and will be aware of the limitations of the polymer matrix materials. They may therefore feel confident in making their choice of plastics suitable to their purpose, especially as carbon fibres are relatively inert to most reagents.

Those engaged in other branches of engineering may need some guidance but it is impossible to be precise and the following notes are given with reserve, to be backed in all cases by environmental testing under realistic conditions over lengthy periods. Frequent service reports should be called for to provide information for data banks. Atmospheric and water pollution can cause severe attack and should be allowed for. The raw edges of materials at joints, cut-outs, and holes may need protection and it must be remembered that the synthetic adhesives used in joints may be attacked.

The advice of the polymer manufacturers, universities, and suitable research laboratories may be sought in making a choice of materials. It must be understood that each of the main groups of polymers covers a wide range of subgroups which may differ in their reactions to chemicals and solvents and that the temperature of the reagent may be a very important factor.

A.3.1 EPOXIDES

Acids: generally unaffected except by high concentrations and wet gases such as chlorine and sulphur dioxide.
Alkalis: unaffected except by wet ammonia, caustic soda, etc.
Solvents: unaffected by alcohols and paraffins. Attacked by chlorinated hydrocarbons and ketones.
Water Absorption: 0·1 per cent
  (Equilibrium)
Temperature Limit: may be up to 250°C. See chapter 2.
Inflammability: moderate to self-extinguishing. Flame retarders may be added but with loss of strength.

A.3.2 POLYESTERS

Acids: fairly resistant except to strong concentrations.
Alkalis: attacked.
Solvents: attacked.
Water Absorption: 0·2 per cent. Crazes at 100°C.
  (Equilibrium)
Temperature Limit: 50°–100°C. See chapter 2.
Inflammability: Poor unless retarders are added but with loss of strength.

252

Acids: attacked by strong acids. May be affected by weak acids.
Alkalis: destroyed by strong alkalis. May be affected by weak alkalis.
Water Absorption: 0·1 per cent upwards.
  (Equilibrium)
Temperature Limit: some types up to 250°C.
Inflammability: good resistance.

A.3.4 POLYIMIDES

Little information yet published. Manufacturers' advice may be sought.
(*Note:* Some of the more commonly used plastics are dealt with in *Chemical Plant Design with Reinforced Plastics*, J. H. Mallinson, McGraw-Hill, New York, 1969.)

A.4 Electro-chemical Corrosion of Metals in Association with Carbon Fibres

Corrosion of metals may occur if they are associated with carbon fibres in the presence of an electrolyte. This may be atmospheric humidity, particularly if it is polluted.

Typical values of the potential differences in volts against a standard electrode are:

Magnesium		+2·34
Zinc		+0·76
Cadmium		+0·71
Aluminium		+0·67
Iron		+0·44
Nickel		+0·25
Tin		+0·14
Lead		+0·13
(Hydrogen		0·00)
CARBON	up to	−0·3
Silver		−0·48
Copper		−0·52
Gold		−0·80

Varying values are quoted by different authorities, probably because of different conditions of test, electrolyte used, purity of the specimens, temperature, etc. Nevertheless the order of values is similar. (*Note:* In American practice the opposite sign convention is used, the cathodic or noble end of the scale being positive and the anodic end negative.)

The maximum potential differences allowed in aircraft work are 0·25 volt for parts normally exposed to the weather and 0·5 volt for interior parts exposed to unpolluted humidity. The degree of attack may be influenced by other factors such as the areas exposed.

It is generally accepted that the polymer matrix provides a barrier, but this does not apply at bolt and rivet holes or if the matrix is cracked or otherwise damaged. At points of contact some organic or inert protection should be provided and inspection should be made possible.

It will be noted that the potential difference between carbon and aluminium is about one volt, and corrosion of the aluminium may therefore be expected. Titanium, not quoted in the table, should be very near to carbon and has very good resistance to attack.

A.5 Thermal Expansion Coefficients of Fibres (along the fibre direction) 10^{-6} per°C

Temp. °C	Type 1	Type 2
− 200 (estimated)	−0·5	−0·03
− 100 (estimated)	−1·3	−0·64
0 (estimated)	−1·65	−0·80
+ 100 (measured)	−1·45	−0·55
+ 200 (measured)	−1·0	−0·15
+ 300 (measured)	−0·4	+0·45

The estimated values are probably accurate to 10 per cent.

A.6 Specific Electrical Resistance of Fibres (along the fibre direction) Micro-ohm-cm

Temp. °C	Type 1	Type 2
25	775	1500
180	660	

Index

262

Printed by J. W. Arrowsmith Ltd., Bristol 3.